Metal Contaminated Aquatic Sediments

Edited by
Herbert E. Allen

Ann Arbor Press

Library of Congress Cataloging-in-Publication Data

Metal contaminated aquatic sediments / edited by Herbert E.
Allen.

p. cm.

Includes bibliographical references and index.
1. Metals--Environmental aspects. 2. Metals--Speciation.
3. Contaminated sediments. I. Allen, Herbert E. (Herbert Ellis),
1939- .
TD879.M47M479 1995
628.1'683--dc20 95-36221
ISBN 1-57504-010-7

ANN ARBOR PRESS, INC.
121 South Main Street, Chelsea, Michigan 48118

PRINTED IN THE UNITED STATES OF AMERICA

To
Ronnie
with appreciation for her help, encouragement and understanding

PREFACE

Over the past decade it has become widely understood that the sediments of rivers, lakes and estuaries in a large number of locations have been contaminated by inorganic and organic materials. These sediments contain a record of past inputs to the aquatic system. A number of processes, such as sorption, tend to immobilize this record while others, such as bioturbation, disturb the record of input. Under certain conditions the contaminants contained in sediments can be released to overlying waters. Thus, sediments may be an important source of contaminants to waters in which littoral and atmospheric contaminants sources have been reduced or eliminated.

Metals are frequent and important contaminants in aquatic sediments. They are subject to a number of reactions in the system including sorption and precipitation and they are greatly influenced by the redox conditions in the sediments. Often the reactions are slow, and reflect biotic processing as well as chemical transformations. These important processes are described in this book. Authors who had intimate knowledge and understanding of these important topics were chosen.

In the initial chapter, Ane de Groot provides a global perspective on the sources and extent of contamination of sediments. He discusses measurement of contaminants and the management of dredged materials. Dr. de Groot was among the first to study contaminated sediments.

The second chapter, by Philippe Van Cappellen and Yifeng Wang, is a discussion of the processes involved in the metal cycling in surficial sediments. They describe a model for the transport-reaction modeling of iron and manganese in sediments.

In the third chapter, George Luther reviews the trace metal chemistry of porewaters. He discusses the sampling methods for obtaining porewaters, the processes which dissolve and precipitate metals, and the types of speciation that occur in porewaters.

Everett Jenne considers the rate of desorption of metals from sediment in the fourth chapter of the book. He shows that to describe the system, both fast and slow components of sorption and desorption reactions must be considered.

In the fifth chapter, Susanne Schultze-Lam, Matilde Urrutia-Mera and Terry Beveridge discuss the composition of bacterial surfaces and how sites on these surfaces interact with metal ions. The importance of bacterial involvement in mineral nucleation and formation is presented and discussed.

Tim Grundl discusses the redox status of sediments in the book's sixth chapter. He discusses the thermodynamic principles that underlie redox measurements and alternative methods to assess the redox status.

The seventh chapter, by John Pardue and William Patrick, shows how alteration of the redox status of sediments results in changes in the speciation of metals in the sediments. Changes in the speciation of arsenic, selenium and chromium and in their solubility, bioavailability and toxicity following alteration of the Eh-pH condition of the sediment are discussed.

John Morse discusses interaction of trace metals with sulfide minerals present in anoxic sediments in the eighth chapter of the book. The interaction of toxic metals with acid volatile sulfide and with pyrite plays a major role in controlling the bioavailability of these metals.

In the ninth chapter the processes of bioturbation are discussed by Gerald Matisoff. Macrobenthos affect the transport of particles and solutes in sediments, modifying the distribution and mobilization of metals.

In the final chapter, Thomas Armitage discusses the U.S. Environmental Protection Agency's strategy for the management of contaminated sediments. He explores strategies to assess, prevent, abate and control sediment contamination as well as strategies for remediation and enforcement and for the management of dredged materials.

This book is the outcome of the Workshop on Metal Speciation and Contamination of Aquatic Sediments which was held in Jekyll Island, Georgia, in June of 1993. This Workshop was the fourth in a biannual series. I want to thank all the authors for their contributions and for their cooperation during the preparation of this book. The expert technical assistance of Ms. Dana M. Crumety, who provided expert secretarial support, is gratefully acknowledged. Finally, I want to thank the Environmental Protection Agency for its financial support of the Workshop.

Herbert E. Allen   Newark, Delaware

Herbert E. Allen is Professor of Civil and Environmental Engineering at the University of Delaware, Newark, Delaware, U.S.A. Dr. Allen received his Ph.D. in Environmental Health Chemistry from the University of Michigan in 1974, his M.S. in Analytical Chemistry from Wayne State University in 1967, and his B.S. in Chemistry from the University of Michigan in 1962. He served on the faculty of the Department of Environmental Engineering at the Illinois Institute of Technology from 1974 to 1983. From 1983 to 1989 he was Professor of Chemistry and Director of the Environmental Studies Institute at Drexel University.

Dr. Allen has published more than 120 papers and chapters in books. His research interests concern the chemistry of trace metals and organics in contaminated and natural environments. He conducted research directed toward the development of standards for metals in soil, sediment and water that take into account metal speciation and bioavailability.

Dr. Allen is past-chairman of the Division of Environmental Chemistry of the American Chemical Society. He has been a frequent advisor to the World Health Organization, the Environmental Protection Agency and industry.

CONTENTS

3   TRACE METAL CHEMISTRY IN POREWATERS

George W. Luther, III

**4   METAL ADSORPTION ONTO AND
DESORPTION FROM SEDIMENTS:
I. RATES**

Everett A. Jenne

Metal Contaminated Aquatic Sediments

METALS AND SEDIMENTS: A GLOBAL PERSPECTIVE

Ane J. de Groot
DLO Research Institute for Agrobiology and Soil Fertility (AB-DLO)
Haren, The Netherlands

1.   INTRODUCTION

In this general view on metal contamination problems of aquatic sediments attention will be paid to the historical development of the pollution, the normalization of measurement results and the behavior of the metals under different environmental conditions.

Furthermore, the influence of large civil engineering projects (such as construction of harbors and enclosure of river outlets) on the behavior of polluted particulate matter will be described, as well as the consequences of high river discharges for the pollution of terrestrial areas. With respect to the bioavailability of the metals, some attention will be paid to the speciation and the quality criteria of the metals. Finally, the management of dredged materials in some countries will be discussed.

For a more detailed treatment of some subjects the reader is referred to the chapters.

2.   DEVELOPMENT OF METAL RESEARCH WITH RESPECT TO POLLUTION

The investigation of heavy metals in stream sediments has become standard practice in mineral exploration since the beginning of the sixties. The concentrations of the metals in river sediments reflect the occurrence and abundance of certain rocks or mineralized deposits in the drainage area of the river [1]. The anthropogenic input of certain metals, however, can equal or exceed the amounts released due to weathering. A first indication of such an input was the anomalous copper concentration in Buzzards Bay, Massachusetts, USA [2]. The excessive amounts of copper came from the industrial center and smelter of New Bedford.

Heavy metal pollution of the Rhine River was first recognized during geochemical investigations of the Rhine sediments in connection with agricultural problems [3]. During the fifties and also thereafter, trace metal research was highlighted in connection with deficiency diseases in agricultural crops. For centuries, embankment of Rhine sediments was an important method to enlarge Dutch agricultural areas. Rhine deposits, however, revealed excessive amounts of copper and zinc, which marked the beginning of very extensive research of river pollution.

Until 1967 the analysis of metals was mainly carried out by classical spectrophotometry and X-ray fluorescence. The investigations of Rhine deposits were highly stimulated by the introduction of activation analysis in 1968 and atomic absorption techniques in 1969. By means of activation analysis Cr, As and a number of rare earth metals were added to the program [4], as well as Hg [5]. Activation analysis, on the other hand, also facilitated estimation of many metals. Special attention was given to the detection of excessive amounts of Cd in Rhine deposits [6, 7].

3.   THE HISTORY OF METAL POLLUTION

Many publications state that the pollution with heavy metals started with the industrial revolution. The industrial revolution, however, is not a univocal concept. In England, this event is placed at 1760-1830, the period in which the country changed from an agricultural to an industrial nation. In other countries, however, the industrial revolution started around 1900, resulting from the rise of electricity, petroleum and automotive industries.

It is, therefore, questionable which period should be selected in this respect. In Figure 1, two examples of Hg-pollution in sedimentary cores of lakes are shown. For Lake Ontario (Canada), the measurements concern materials deposited since about 1900 [8]. With some imagination the curve of Hg contents might approach the background value as it existed around 1850. Similar reasoning is possible for the Hg-contents in sediments from Lake Windermere (England). The measurements in this lake concern deposits as far back as about 1870 [9].

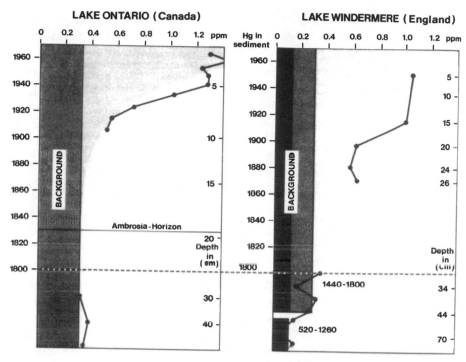

Figure 1. *Mercury in Sedimentary Core Profiles from Lake Ontario (Thomas [8]) and Lake Windermere (data from Aston et al. [9]).*

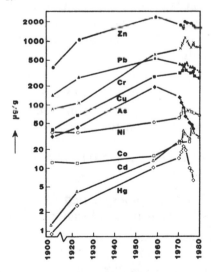

Figure 2. *History of Trace Metal Pollution in the Rhine River in The Netherlands as Reflected in its Sediments (after Salomons et al. [10]).*

Another example is the history of metal pollution in sediments from the Rhine River, as shown in Figure 2 [10]. The pollution of these deposits goes back at least as far as 1900.

The latest data in Figure 2 are from 1980. Between 1920 and 1960 all metal concentrations determined in Rhine River sediments have increased; Cd, Cu and Cr continued to increase between 1960 and 1975. The concentration of Pb, Zn, Hg and As decreased. However at present, most metal contents are close to the levels of 1900.

4.   NORMALIZATION OF MEASUREMENT RESULTS

4.1   Extrapolation of regression curves

Metal contents of sediments strongly depend on the granular composition of the deposits. Due to a preferred occurrence of the heavy metals in the finest grain-size fractions, linear relationships are generally found between the heavy-metal contents and the fraction of particles <20 μm (expressed as a percentage of the $CaCO_3$-free mineral constituents) in samples from the same location. These linear relationships make it possible to characterize the contents of a specific metal of a whole group of co-genetic sediments by a single value, the content being obtained through extrapolation of the regression line to 100% of the fraction < 20 μm. A more reliable method, however, is the estimation of the metal content at 50% of the fraction < 20 μm, which corresponds better to the natural granular conditions.

The calculation of a regression line, however, requires a large number (10-15) of samples from one location. Moreover, it is often impossible to determine a regression line, due to the limited range in particle size of the sediment at a given site.

4.2   Analysis of a relevant particle-size fraction

Separation of particles of a specified size is advantageous, because only a few samples from a particular location are needed.

Heavy metals are present in considerable amounts in the fractions up to 20 μm and probably up to 35 μm [11-13]. The use of the fraction < 63 μm for metal analysis is internationally generally accepted (e.g. ICES, International Council for the Exploration of the Sea) and is preferred to the analysis of smaller fractions for several reasons: (a) amounts of the heavy metals, (b) efficiency of sieving procedure, and also (c) the fact that this fraction corresponds to the material that is transported in suspension over large distances with river and sea currents.

A special advantage of the analysis for metals in a separated fraction concerns the analysis of very sandy deposits. When the contents of contaminants in these deposits are determined in the entire sample, they will hardly differ from those in a similar non-contaminated sample. The contaminants in the fine-grained fraction, however, may affect those biota which selectively take their food from the fine fractions.

4.3   Geochemical normalization

Geochemical normalization is the estimation of the ratio between the contents of the various heavy metals and the content of a conservative element, such as Al, Sc or Li. The last elements are mainly present in the fine-grained fractions.

Within estuaries, especially Sc shows a very conservative behavior. Ackermann [14] efficiently used Sc as a reference element in his studies on heavy metal behavior in the Ems estuary. A disadvantage of Sc, however, is the fact that this element is usually determined by neutron activation analysis. Loring [15] concluded that normalization by means of Al is not applicable to sediments from higher latitudes, because the sand fractions of sediments derived from glacial erosion contain much Al. This author emphasizes the use of Li. This element is conservative in its behavior and can easily be detected by atomic absorption spectrometry.

4.4   General remark

It must be emphasized that the methods for the normalization both of inorganic elements and organic compounds are still under discussion. Especially the ICES pays attention to the applicability of normalization methods over wide geographical areas and a range of sediment types.

5.   MOBILIZATION OF METALS

As a consequence of changing environmental conditions, heavy metals can be released from the sediment. This generally happens during transport of particulate matter through estuaries and as a consequence of erosion and redeposition of sediments. Suddenly occurring intensive mobilization processes will be marked as chemical time bombs.

In estuaries, salinity increases into a seaward direction. Although the physical and chemical processes in the mixing zone are complicated with respect to certain metals, especially Cd, intensive mobilization from riverine particles takes place between 1% and 6% salinity. The formation of complexes with Cl⁻ ions is evident in this connection. The influence of Cl⁻ ions, however, is not the only cause of mobilization processes. Already in the upper part of an estuary (less than 0.5% salinity), loss of metals can take place as a consequence of decomposition of particulate organic matter. This has been clearly demonstrated by Jouanneau *et al.* [16, 17] for the element Zn in the Gironde estuary. Details about the character of these processes should be further investigated.

Mobilization of metals as a consequence of erosion and redeposition of sediments is shown in Figure 3. Under anoxic conditions the metals are immobilized as sulfides. After erosion and oxidation of the material, the metals are transformed into unstable forms and, especially under marine conditions, are easily released and complexed with chloride ions.

A chemical time bomb (CTB) is a chain of events resulting in the delayed and sudden occurrence of harmful effects due to mobilization of chemicals stored in sediments as a consequence of alteration of certain environmental conditions (Stigliani *et al.* [18]).

Important soil properties controlling the release of heavy metals from sediments are mainly pH, redox potential, salinity and probably biomethylation processes.

Drainage of contaminated coastal wetlands in brackish environments can give rise to serious problems. If these soils are drained, pyrite is oxidized and the sulfuric acid as a consequence of this reaction can cause a drop in pH below 4.5 as soon as the buffer mechanisms are depleted. Heavy metals will show a non-linear increase under these conditions and are suggestive of a "chemical time bomb".

A striking example of the danger of a chemical time bomb in The Netherlands are the deposits of the River Meuse. Extremely high Cd and Pb contents are accompanied by low contents of carbonates. The hydrological conditions in the area of the river mouth (easy removal of calcium bicarbonate) will cause a considerable decalcification and drop in pH, and consequently mobilization of the heavy metals within a few decades.

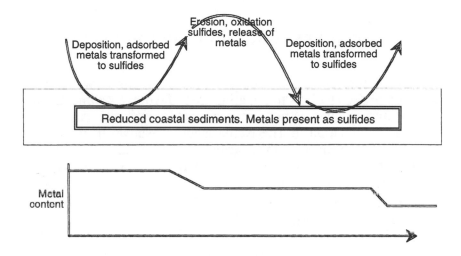

Figure 3.   *Decrease in Metal Concentrations in Coastal Sediments through Cyclic Processes of Erosion and Deposition.*

Another example of the fast mobilization of metals is the resuspension of contaminated anaerobic dredged materials, especially under marine conditions.   Oxidation of the dredged material and subsequent solubilization of the metals (exchange for Na^+ ions and complexation by Cl^- ions) can cause serious problems.

Stigliani [19] refers to the influence of sea level rise on the increase in salinity of vulnerable estuaries, which alters the balance between fresh and salt waters. Since toxic materials often preferentially accumulate in the sediments of estuarine regions, the danger of enhanced landward salinity exists, resulting in mobilization of toxic materials in highly polluted estuaries.

Still open to question is to what extent biomethylation of metals can function as a chemical time bomb [20, 21]. For instance, metallic mercury can be transformed into the very toxic methyl-mercury through the action of bacteria. Environments favorable to this conversion are warm aquatic areas rich in organic matter, slightly acidic and with a high bacterial activity. A region of great concern in this connection is the Amazon area in South America. Mercury is increasingly used in these areas as a means to amalgate gold in the process of gold mining. Huge amounts of Hg accumulate in the Amazon ecosystem via the atmosphere. The existence of a chemical time bomb in this area has to be substantiated with more research.

6.   CIVIL ENGINEERING AND METAL PROBLEMS

The enclosures of river mouths and coastal lagoons, as well as the construction of deep harbors in estuarine areas, favor the sedimentation of polluted material and restrict discharges out into the sea. A striking example is the situation in The Netherlands, as shown in Figure 4. Outlets of the Rhine River are the distributary Yssel (which enters Lake Yssel), the Rotterdam harbor area and the Haringvliet. Lake Yssel was created by enclosure of the former Zuyder Sea, the Haringvliet area was enclosed as part of the well-known delta plan [22].

In Lake Yssel the enrichment of the bottom with metals is not only caused by a more intensive sedimentation of suspended matter; in the lake a part of the carbon dioxide is withdrawn from the water phase as a consequence of escape to the air, as well as consumption of the carbon dioxide by algal blooms (intensive eutrophication of the Rhine water). Withdrawal of carbon dioxide causes a rise in pH of the water from 7.4 to 9.0. Over this range the adsorption of metals such as Zn, Cr and Cd increases considerably. The adsorption of metals to suspended matter in Lake Yssel is furthermore favored by its shallowness (easy erosion of the sediments due to wind force). The water leaving the lake to the North Sea contains low amounts of metals.

The heavy-metal enrichment of the bottom of the much deeper Haringvliet is mainly caused by enhanced sedimentation. This is also the case with the deep harbors of Rotterdam.

The sedimentation of polluted sediments in the three areas mentioned causes serious environmental problems.

7.   TERRESTRIAL CONSEQUENCES OF METAL
        POLLUTION

During high water discharge of rivers, generally in Spring, the banks of the river can be flooded and covered by a layer of (polluted) sediment. Figure 5 shows a cross section of a typical floodplain-river system in the lower Rhine area.

Figure 4.   *The Rhine River and its Distributaries in The Netherlands.*

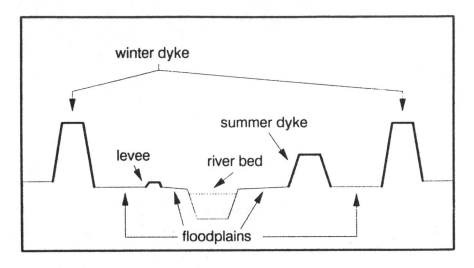

Figure 5.   *Cross Section of a Typical Dyke-protected Floodplain*
               System in the Dutch Lowland.

Generally, a low summer dyke and a high winter dyke are part of the system. At some locations, however, no summer dyke is present (left-hand side of Figure 5).

It appears that the sediments deposited on the flood plains are less polluted than the material carried by the river under normal conditions [23]. Some examples:

- Sediment deposited under normal flow conditions 1855 µg Zn/g, 27 µg Cd/g.

- Flood plain sediment without summer dykes 943 µg Zn/g, 12 µg Cd/g.

- Flood plain sediment with summer dykes 690 µg Zn/g, 8 µg Cd/g.

Obviously the river carries a different kind of material during periods of high water discharge. A possible explanation is the erosion of the edges and the bottom of the river under these conditions. The effect of the water discharge on the metal content is also demonstrated by the differences in metal content with and without summer dykes.

In The Netherlands a set of values was developed for contaminant levels in agricultural soils. These values are intended to indicate the levels above which problems might arise for plant and animal well-being, or for the quality of agricultural products. Flood plain sediment deposits can be considered as clay soils used for grazing cattle and sheep. The maximum permissible contents of Zn and Cd are 350 and 3 $\mu g/g$, respectively. The levels of these elements found in most samples from the river flood plains were well above these values.

8.   SPECIATION AND QUALITY CRITERIA

Although total content of metals can give insight into the degree of pollution, it gives no information on bioavailability, nor on the chemical and physical behavior under the relevant environmental conditions of the sediment.

8.1   Single extractions

In an extensive literature review Campbell et al. [24] discussed extraction solutions for the estimation of heavy-metal bioavailability. They conclude that the various species play an important role in the process of metal uptake and therefore in the extraction medium to be selected. This means that the extraction medium has to be adapted to those species which are available for the biota. Attention has been focused on benthic invertebrates, especially in the estuarine and marine environment. Metal uptake under these conditions was more closely correlated with easily extractable fractions, e.g. with 3M HCl, than with total metal contents.

The uptake of many metals from submerged soils by macrophytes (*Spartina* species, *Cyperus esculentus*) was satisfactorily reflected in extraction with DTPA in combination with $CaCl_2$ and TEA (Lee et al. [25, 26]).

8.2   Sequential extractions

With respect to the speciation of metals in aerobic sediments sequential extraction procedures were first carried out by Gibbs [27] on suspended materials from the Amazon and Yukon Rivers. Engler et al. [28], prepared a practical selective extraction procedure for sediment characterization ("elemental partition").

Commonly known is the system of analysis developed by Tessier et al. [29], in which the heavy metals are classified according to their association with different phases of the sediment. Through sequential extractions they determined exchangeable metals, metals associated with carbonates, Mn-oxides, Fe-oxides, organic matter, as well as metals incorporated in crystal lattices.

Although the authors characterize their method as "operational" in nature, they believe that the method provides useful comparative information about regional variations in sediment chemistry.

Jenne *et al.* [30] used this method as a starting point for modelling the bioavailability of heavy metals in aerobic fresh-water environments. The metals sorbed by three of the earlier mentioned phases are considered to be in equilibrium with the bioavailable metal ions in the surrounding solution: metals sorbed by the oxides of manganese and iron and by the organic matter. The metal ion concentration is expressed by:

$$\left[Me^{2+} \right] = \frac{Me_s}{k_{Fe}[FeOx] + K_{Mn}[MnOx] + K_{POC}[Org\ C]} \tag{1}$$

The metal ion concentration is given here as a function of easily measurable quantities. Me_s is the total amount of the relevant metal in the extracts of the three phases; FeOx, MnOx and Org C are the phases themselves. The equilibrium constants K_{Fe}, K_{Mn} and K_{POC} are estimated in separate experiments. It appears that there are linear relationships between log K-values and pH.

Much work is done at this moment to develop sediment quality criteria for metals with the above-mentioned "equilibrium partitioning approach".

The method seems to be promising for the prediction of metal contents in biota living on or in the sediments (*Macoma balthica, Elliptio complanata, Nereis diversicolor*).

8.3   Characterization of anaerobic sediments

Under anaerobic conditions the bioavailability of heavy metals is largely controlled by the absorption and coprecipitation of the metals with sulfide minerals.

Di Toro *et al.* [31] give a comprehensive description of the role of sulfide in the chemical activity of the metal in the sediment-interstitial water system. The system is characterized by treatment of the sediment with cold hydrochloric acid. The sulfide fraction released in this way is referred to as the acid volatile sulfide or AVS. The metal concentration that is simultaneously extracted is termed simultaneously extracted metal or SEM. Characteristic of the mobility of the metals is now the ratio SEM/AVS, expressed as the molar sum of the metals (Cd, Cu, Hg, Ni, Pb, Zn) and the molar concentration of acid volatile sulfide. For SEM/AVS < 1, no acute toxicity has been observed in any sediment for any benthic organism.   If the ratio is > 1, the mortality of sensitive species increases in the range of 1.5-2.5 µmol of SEM/µmol of AVS.

In Chapter 8, J.W. Morse describes the processes after resuspension of anaerobic sediments, seasonal migration of the redox-cline and dumping of dredged material on land.   It has been found that 20 to > 90% of the pyrite-bound metals can be released in a day or less by exposure to oxic seawater and become potentially bioavailable.

9.   MANAGEMENT OF DREDGED MATERIALS

World-wide, annually 600 million tonnes of dredged materials have to be deposited somewhere.   One-fourth of all materials is ocean-dumped and another two-thirds is deposited in wetlands and nearshore [32].   A considerable part of the dredged materials contains excessive amounts of pollutants.   Each country has its own regulations and techniques to handle the contaminated materials.   We mention here a few details about the management of the dredged materials in the U.S., The Netherlands and Germany.

In the U.S. the Corps of Engineers is responsible for the dredging operations of harbors and waterways.   About 10 percent of the materials is contaminated.   The Waterways Experiment Station in Vicksburg, Mississippi, has developed a tiered approach to assess the quality of the sediments.   In Tier I existing information is gathered about the relevant sediments.   If this knowledge is not enough to decide about the disposal of the sediment, in Tier II physical and chemical analyses are carried out.   If necessary, the investigations can be continued with Tier III (bioassays in the laboratory) and Tier IV (bioaccumulation and long-term tests).   All tests take both the water column and the sediment deposit into account.

Subsequently, the disposal management strategy is considered. Alternatives are upland disposal (e.g. long-term confined, habitat development), intertidal sites (unconfined on mudflats, confined by boundary structures) or subaqueous disposal (unconfined, capped in an excavated pit).

In The Netherlands, sediments of the Rhine delta are heavily contaminated.   Dredged materials from the Rotterdam harbor areas are divided into four classes, varying from almost clean to heavily polluted. The ultimate goal with respect to the pollution problem of the Rhine is that in the year 2010 the particulate matter of this river will fulfill the requirements of clean sediment.   At this moment class 2 and 3 materials (light and moderately polluted) are deposited in a containment area, which is located as a peninsula in the mouthing area of the Rotterdam harbor. This "sludge island" is surrounded by a high-tide resistant ring wall. Model calculations suggest that the concentration of most contaminants will not affect groundwater composition significantly. For the disposal of heavily contaminated, class 4, dredged material a special site is prepared in such a way that the dredged material is placed above the highest groundwater level. To protect the quality of the groundwater, the dredged material is packed in a liner.

The policy for the dredging problems in The Netherlands in the near future is to treat 20% of the materials (e.g., hydrocyclonage of relatively sandy mud). For the rest of the polluted materials, containment facilities will have to be constructed.

Although containment above the groundwater level is preferred from the viewpoint of soil protection, only a few small local facilities of this kind will be constructed. The further intentions are to create two large confined disposal areas within fresh water lakes.

Two possibilities are shown in Figure 6: a subaqueous disposal site and an island site with containment capacities of 25×10^6 m^3 per unit. Both constructions need an effective insulation. Among the convential liners, high density polyethylene and polypropylene seem to be the most useful.

We mention here, however, two other possibilities: the use of precipitation layers and the application of polyelectrolytes. The principle of the precipitation layer is shown in Figure 7 [33].

Two adjacent sand layers are mixed with $Ca(OH)_2$ and $MgCO_3$, respectively. Both compounds are slightly soluble. At the boundary of the two layers the two compounds react, forming an impermeable layer of $CaCO_3$ and $Mg(OH)_2$.

Another possibility is the use of high-molecular polyelectrolytes. These compounds can be synthesized in such a way that they are fixed by the clay particles, are able to chelate diffusing heavy metals and at the same time react with PAH's and PCB's.

In Figure 8 the complexation of Cu-ions by a polymer molecule is shown. The linear chain contains styrene units, interrupted by amine ligands. Underneath the polymer-chain a part of the chain is drawn with a 4-vinylpyridine unit as an amine ligand.

The polyelectrolytes are very effective in preventing the metal ions from escaping by two different mechanisms: the polydentate effect (several ligands surrounding the metal) and the polymer chain effect (flexibility of the chain). The compounds can be used in very low concentrations, without destroying the structure of the sediment.

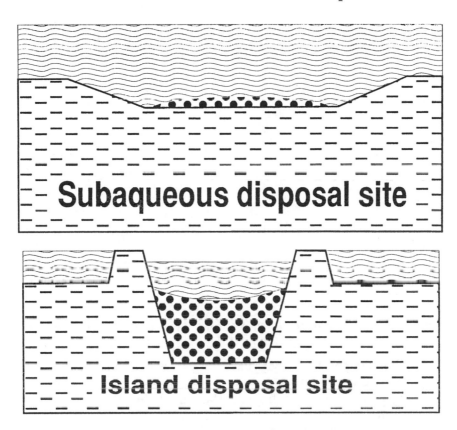

Figure 6.   *Disposal Sites for Dredged Materials.*

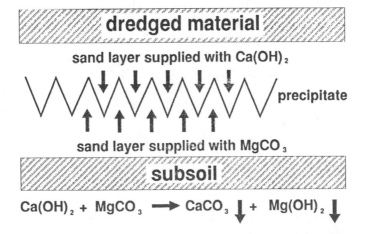

Figure 7.   *Building a Precipitation Layer as a Liner for Dredged Materials.*

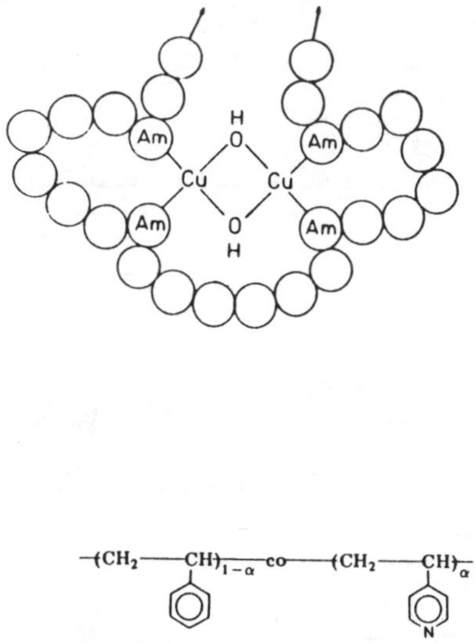

Figure 8.   *Complexation of Cu Ions by a Polymer Molecule (co-polymer of styrene and 4-vinylpyridine) [34].*

Although polyelectrolytes have not yet been used in practice, their use in restricting the dispersion of pollutants is very promising. In Germany polluted sediment from the Hamburg harbor area is treated in a large-scale operation with a highly effective combination of hydrocyclonage and elutriator [32]. In the hydrocyclone the separation of the course fraction (relatively clean sand) from the highly polluted fines is effected by the action of centrifugal forces. The coarse fraction leaves the cyclone in the underflow, while the fine fractions are contained in the overflow. The sharpness of this separation is fairly low; therefore, the separation is completed by leading the coarse fraction through an elutriator. The basic principle here is separation according to the settling velocity of the particles in an upflowing water stream. The sand thus separated has a heavy-metal content which equals that occurring under natural conditions. The polluted fines are stored in a confined disposal site.

REFERENCES

1.   Hawkes, H.E. and J.S. Webb, *Geochemistry in Mineral Exploration* Harper & Row, (New York, 1962).

2.   Moore, J.R., "Bottom Sediment Studies, Buzzard Bay, Mass," *J. Sediment. Petrol.* 33:511-558 (1963).

3.   de Groot, A.J., "Mobility of Trace Elements in Deltas," in *Soil Chemistry and Fertility* G-V Jacks, Ed., pp. 267-279 Trans. Comm. II & IV ISSS, (Aberdeen, 1966).

4.   de Groot, A.J., K.H. Zschuppe, M. de Bruin, J.P.W. Houtman and P. Amin Singgih, "Activation Analysis Applied to Sediments from Various River Deltas," in *The 1968 International Conference on Modern Trends in Activation Analysis*, pp. 62-71 (Gaithersburg, USA 1968).

5.   de Groot, A.J., J.J.M. de Goeij and C. Zegers, "Contents and Behaviour of Mercury as Compared with Other Heavy Metals in Sediments from the Rivers Rhine and Ems," *Geologie en Mijnbouw* 50:393-398 (1971).

6.   de Groot, A.J., "Occurrence and Behaviour of Heavy Metals in River Deltas, with Special Reference to the Rhine and Ems Rivers," in *North Sea Science*, E.D. Goldberg, Ed., pp. 308-325 MIT Press, (Cambridge, 1973).

7.   de Groot, A.J., "De samenstelling van het slib in het Rotterdamse haven-gebied," *Mededelingen van de KNHM* 50:11-14 (1974).

8.   Thomas, R.L., "The Distribution of Mercury in the Sediment of Lake Ontario," *Can. J. Earth Sci.* 9:636-651 (1972).

9.   Aston, S.R., D. Bruty, R. Chester and R. Padgham, "Mercury in Lake Sediments. A Possible Indicator of Technological Growth," *Nature* (London) 24:450-451 (1973).

10.   Salomons, W., W. van Driel, H.Kerdijk and R. Boxma, "Help! Holland is Plated by the Rhine (environmental problems associated with contaminated sediments)," Effects of Waste Disposal on Groundwater. *Proc. Exeter Symp. (HAS)* 39:255-269 (1982).

11.   Ackermann, F., N. Bergmann and U. Schleichert, "Monitoring of Heavy Metals in Coastal and Estuarine Sediments - A Question of Grain-size: < 20 versus < 60 µm," *Environ. Technol. Lett.* 4:317-328 (1983).

12.   Schoer, J.H. "Eisen-oxo-hydroxide und ihre Bedeutung für die Schwermetall-verteilung in Ästuarien, insbesondere der Elbe," *Thesis*, pp. 240-244 (Heidelberg, 1984).

13. de Groot, A.J., "Manganese Status of Dutch and German Holocene Sediments," *Versl. Landbkd. Onderz.* 69.7:37-41 (1963).

14. Ackermann, F., "A Procedure for Correcting Grain Size Effect in Heavy Metal Analysis of Estuarine and Coastal Sediments," *Environ. Technol. Lett.* 1:518-527 (1980).

15. Loring, D.H., "Normalization of Heavy Metal Data," *Rep. ICES Working Group on marine sediments in relation to pollution*, (1988).

16. Jouanneau, J.M., H. Etcheber and C. Latouche, "Impoverishment and Decrease of Metallic Elements Associated with Suspended Matter in the Gironde Estuary," in *Trace Metals in Sea Water*, C.S. Wong, E. Boyle, K.W. Bruland, J.D. Burton and E.D. Goldberg, Eds., pp. 245-263 Plenum Publ. Corp., (1983).

17. Jouanneau, J.M., C. Latouche and H. Etcheber, "Les flux de Zn, Pb, Cu et du carbone organique à l'océan, exportés par la Gironde," *Rapp. V. Réun. Cons. Int. Explor. Mer*, 186:289-300 (1986).

18. Stigliani, W.M., P. Doelman, W. Salomons, R. Schulin, G.R.B. Smidt and S.E.A.T.M. van der Zee, "Chemical Time Bombs: Predicting the Unpredictable," *Environment* 33:4-30 (1991).

19. Stigliani, W.M., "Overview of the Chemical Time Bomb Problem in Europe," in *Chemical Time Bombs. Proc. European State-of-the-Art Conference on Delayed Effects of Chemicals in Soils and Sediments*, G.R.B. ter Meulen, W.M. Stigliani, W. Salomons, E.M. Bridge and A.C. Imeson, Eds., pp. 13-29 Foundation for Ecodevelopment, Hoofddorp, (The Netherlands, 1993).

20. Lacerda, L.D. and W. Salomons, "Mercury in the Amazon: A Chemical Time Bomb?" *Rep. Dutch Ministry of Housing, Physical Planning and the Environment*, (1991).

21. Stigliani, W.M. and W. Salomons, "Our Fathers' Toxic Sins," *New Scientist*, 140:38-42 (1993).

22. de Groot, A.J. and W. Salomons, "Influence of Civil Engineering Projects on Water Quality in Deltaic Regions," in *Effects of Urbanization and Industrialization on the Hydrological Regime and on Water Quality,* IAHS 123:351-357 (1977).

23. Japenga, J., K.H. Zschuppe, A.J. de Groot and W. Salomons, "Heavy Metals and Organic Micropollutants in Floodplains of the River Waal, A Distributary of the River Rhine 1958-1981," *Neth. J. Agric. Sci.* 38:381-397 (1990).

24. Campbell, P.G.C., A.G. Lewis, P.M. Chapman, A.A. Chowder, W.K. Fletcher, B. Imber, S.N. Luoma, P.M. Stokes and M. Winfrey, "Biologically Available Metals in Sediments," *Nat. Res. Counc. Can.*, publ. no. NRCC 276694, pp. 135-148 (1988).

25. Lee, C.R., R.M. Smart, T.C. Sturgis, R.N. Gordon and M.C. Landin, "Prediction of Heavy Metal Uptake by Marsh Plants Based on Chemical Extraction of Heavy Metals from Dredged Material," *U.S. Army WES, Tech. Rep.* D-78-6, (1978).

26. Lee, C.R., B.L. Folsom and R.M. Engler, "Availability and Plant Uptake of Heavy Metals from Contaminated Dredged Material Placed in Flooded and Upland Disposal Environments," *Environ. Intern.* 7:65-71 (1982).

27. Gibbs, R.J., "Mechanisms of Trace Metal Transport in Rivers," *Science* 180:71-73 (1973).

28. Engler, R.M., J.M. Brannon, J. Rose and G. Bigham, "A Practical Selective Extraction Procedure for Sediment Characterization," in *Chemistry of Marine Sediments*, T.F. Yen, Ed., pp. 163-171 Ann Arbor Science, (Ann Arbor, 1977).

29. Tessier, A., P.G.C. Campbell and M. Bisson, "Sequential Extraction Procedure for the Speciation of Particulate Trace Metals," *Anal. Chem.* 51:844-851 (1979).

30. Jenne, E.A., D. DiToro, H. E. Allen and C. Zarba, "An Activity-Based Model for Developing Sediment Criteria for Metals: I. A New Approach," *Proc. Int. Conference: Chemicals in the Environment, Lisbon, Portugal July 1-3*, pp. 560-568 (1986).

31. Di Toro, D.M., J.D. Mahony, D.J. Hansen, K.J. Scott, A.R. Carlson and G.T. Ankley, "Acid Volatile Sulfide Predicts the Acute Toxicity of Cadmium and Nickel in Sediments," *Environ. Sci. Technol.* 26:96-101 (1992).

32. Förstner, U., "Contaminated Sediments," *Lecture Notes in Earth Sciences 21*. Springer-Verlag, (Berlin, 1989).

33. Van der Sloot, H.A. and P.L. Côté, "Modelling Interactions at a Waste/Waste Interface," *Environ. Technol. Lett.* 6:359-368 (1985).

34. Challa, G., "Polymer Chain Effects in Polymeric Catalysis," *J. Mol. Catal.* 21:1-16 (1983).

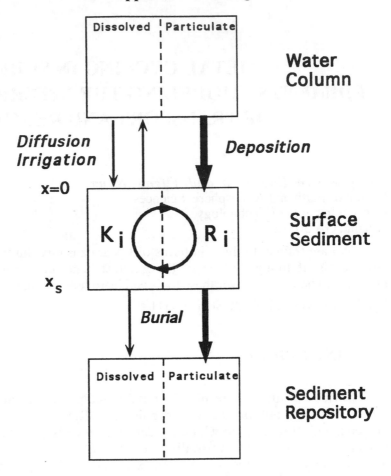

Figure 1.   *Schematic Representation of Reactions and Fluxes Regulating the Fate of Metal Species in Surface Sediments. The boundary conditions at the water-sediment interface are the bottom water composition and the particulate deposition fluxes. Within the surface sediment, the distributions of the species are modified by irreversible reactions and rapid, reversible equilibria. Solute species undergo transport via molecular diffusion, sediment mixing and irrigation. Solid species are transported by sediment accumulation and particle reworking. The coupled transport and reaction processes result in the net uptake or release of dissolved species at the water-sediment interface and in the permanent burial of particulate and dissolved species in the sediment column.*

Mathematical sediment models quantitatively describe mass transfer of metals resulting from transport and reaction. These models can be used to test the sensitivity of the system to the various transport and reaction parameters, as well as to variations in particulate deposition fluxes and bottom water composition. When properly parameterized, they allow the prediction of the response of sediment-water exchanges, porewater quality, and particulate burial fluxes to changing conditions in the overlying aquatic environment.

This paper outlines the mathematical theory of transport-reaction modeling as applied to metals in surface sediments. Based on the general theory, a model for iron and manganese in sediments is developed, and the cycling of the two metals is simulated in a number of representative depositional environments.

2. SURFACE SEDIMENTS

The most intense biogeochemical activity is concentrated within the surface layer of a sediment. This is also the portion of the sediment that directly affects the quality of the overlying water body through solute exchanges (Figure 1). The water-sediment interface constitutes the upper boundary of the surface sediment. It separates the water column, with its large-scale fluid motion and turbulent mixing, from the sediment where molecular diffusion typically dominates solute transport. For particulates, the transfer across the water-sediment interface corresponds to a transition from a relatively rapid sinking rate to a much slower sediment accumulation rate. Thus, compared to the water column, a surface sediment may be viewed as a semi-confined environment characterized by a long residence time of the particulate matter. These conditions favor the extensive biogeochemical processing of the deposited materials and the establishment of steep compositional porewater gradients.

Localizing the lower boundary of a surface sediment is less straightforward. Here, it is proposed to define this boundary as the maximum depth in the sediment at which biogeochemical reactions still exert a significant effect on solute exchanges at the water-sediment interface. Rough estimates of this depth can be obtained from a dimensional analysis of reaction and transport parameters.

Consider a reaction taking place in a given sediment layer at depth L below the water-sediment interface. Ignoring enhanced solute transport by macrofaunal activity or wave pumping, a solute produced during the reaction may still reach the water-sediment interface if L $\leq$

D_s/ω, where D_s is the bulk sediment molecular diffusion coefficient of the dissolved species (on the order of 100 cm^2 a^{-1}) and ω is the linear sediment accumulation rate. In slowly depositing sediments, such as deep-sea sediments ($\omega \approx 10^{-4}$-10^{-3} cm a^{-1}), the ratio D_s/ω overestimates the thickness of the biogeochemically active layer of the sediment. In those cases, better estimates of the lower boundary of the active surface sediment are provided by the ratios ω/k or $(D_{mix}/k)^{1/2}$, where k is the pseudo-first order rate constant characterizing the "dominant" (bio)chemical reaction, usually identified as the oxidation of organic matter (see below), and D_{mix} is the sediment mixing coefficient (see section 4.4). The ratio ω/k applies to a surface sediment where advection dominates particle transport; $(D_{mix}/k)^{1/2}$ should be used when mixing dominates.

Sediment accumulation rates, sediment mixing coefficients and organic carbon oxidation rate constants vary over several orders of magnitude in marine and freshwater depositional environments [2,3]. Nonetheless, the characteristic length scales predicted by the ratios D_s/ω, ω/k or $(D_{mix}/k)^{1/2}$ - whichever applies - fall mostly within the range 10-100 centimeters. Thus, particulate matter that is buried at depths exceeding one meter may be considered to be permanently stored into the sediment repository, i.e., its further transformation no longer affects the overlying aquatic environment.

Most chemical transformations affecting metals in surface sediments are driven, directly or indirectly, by the decomposition of organic detritus deposited from the water column. Table 1 illustrates some of the changes in porewater and solid sediment chemistry brought about by organic matter degradation. The primary reactions A-1 to A-6 describe the net oxidation of organic matter, while A-7 to A-20 are secondary reactions involving the products formed during the primary oxidation reactions. The spatial and temporal distributions of these reactions control the cycling of metals in sediments.

Degradation of organic detritus deposited from the water column is the ultimate source of energy for the microbial and macrofaunal populations that inhabit the surface sediment. This "external" supply of metabolic energy maintains the overall state of chemical disequilibrium of the surface sediment. This is expressed, among other things, by the persistence of vertical gradients in chemical and biological properties in surface sediments.

Because surface sediments are open, non-equilibrium systems, a kinetic description which accounts for reaction and transport is required. Under special circumstances time-invariance may be achieved, but it will correspond to a dynamic steady-state rather than true thermodynamic equilibrium. Local equilibrium may be invoked for reactions that are fast relative to the characteristic time scales of the transport processes. For instance, solution and surface speciation reactions are usually assumed to reach equilibrium. In general, however, the task at hand will be to provide kinetic expressions for the irreversible reactions and mass fluxes that take place in the sediment.

3.   MATHEMATICAL AND NUMERICAL THEORY

Mass conservation underlies the quantitative description of sediment biogeochemistry. In its most general form, the conservation of a porewater or solid sediment constituent is given by the following partial differential equation [2,4]:

$$\frac{\partial \hat{C}}{\partial t} = \sum \nabla \bullet \hat{J} + \sum \hat{S} + \sum \hat{R} \tag{1}$$

where t is time, $\hat{C}$ is the concentration of the constituent per unit volume of total sediment, $\nabla \bullet \hat{J}$ is the divergence of the local transport flux $\hat{J}$ affecting the constituent (e.g., sediment advection, porewater diffusion, local sediment mixing), $\hat{S}$ is a source or sink of the constituent resulting from nonlocal transport (e.g., irrigation and large scale sediment reworking), $\hat{R}$ is the rate of a biogeochemical transformation (e.g., chemical reaction, microbial production or consumption, radioactive decay).

Most data sets on sediment and porewater properties are in the form of depth profiles. Equation 1 is therefore simplified to the one-dimensional vertical case by assuming homogeneous horizontal distributions of species; and we obtain the following second-order partial differential equation:

$$\frac{\partial \zeta C}{\partial t} = \frac{\partial}{\partial x}\left\{ D\zeta \frac{\partial C}{\partial x} - \omega \zeta C \right\} + \zeta \sum S + \zeta \sum R \tag{2}$$

where x is depth below the water-sediment interface, D is the total dispersion-diffusion coefficient (in units of surface area sediment per unit of time), and ω is the vertical sediment advection rate (in units of length sediment per unit of time). Equation 2 can be used to describe

Table 1. Irreversible Reactions (A-1 to A-19) and Alkalinity Conservation (A-20). Reactions A-1 to A-6 represent the net degradation of organic matter deposited from the water column. Reactions A-7 to A-16 describe the reoxidation of secondary species produced during the oxidation of organic matter. Reactions A-17 to A-19 correspond to the non-reductive precipitation of carbonate and sulfide mineral phases. The irreversible production or consumption of protons is buffered by the dissolved carbonate/sulfide acid-base interconversions (A-20).

$$(CH_2O)_x(NH_3)_y(H_3PO_4)_z + (x+2y)O_2 + (y+2z)HCO_3^- \xrightarrow{R_1} (x+y+2z)CO_2 + yNO_3^- + zHPO_4^{2-} + (x+2y+2z)H_2O \quad \text{A-1}$$

$$(CH_2O)_x(NH_3)_y(H_3PO_4)_z + \left(\frac{4x+3y}{5}\right)NO_3^- \xrightarrow{R_2} \left(\frac{2x+4y}{5}\right)N_2 + \left(\frac{x-3y+10z}{5}\right)CO_2 + \left(\frac{4x+3y-10z}{5}\right)HCO_3^- + zHPO_4^{2-} + \left(\frac{3x+6y+10z}{5}\right)H_2O \quad \text{A-2}$$

$$(CH_2O)_x(NH_3)_y(H_3PO_4)_z + 2xMnO_2 + (3x+y-2z)CO_2 + (x+y-2z)H_2O \xrightarrow{R_3} 2xMn^{2+} + (4x+y-2z)HCO_3^- + yNH_4^+ + zHPO_4^{2-} \quad \text{A-3}$$

$$(CH_2O)_x(NH_3)_y(H_3PO_4)_z + 4xFe(OH)_3 + (7x+y-2z)CO_2 \xrightarrow{R_4} 4xFe^{2+} + (8x+y-2z)HCO_3^- + yNH_4^+ + zHPO_4^{2-} + (3x-y+2z)H_2O \quad \text{A-4}$$

$$(CH_2O)_x(NH_3)_y(H_3PO_4)_z + \frac{x}{2}SO_4^{2-} + (y-2z)CO_2 + (y-2z)H_2O \xrightarrow{R_5} \frac{x}{2}H_2S + (x+y-2z)HCO_3^- + yNH_4^+ + zHPO_4^{2-} \quad \text{A-5}$$

$$(CH_2O)_x(NH_3)_y(H_3PO_4)_z + (y-2z)H_2O \xrightarrow{R_6} \frac{x}{2}CH_4 + \left(\frac{x-2y+4z}{2}\right)CO_2 + (y-2z)HCO_3^- + yNH_4^+ + zHPO_4^{2-} \quad \text{A-6}$$

$$Mn^{2+} + \frac{1}{2}O_2 + 2HCO_3^- \xrightarrow{R_7} MnO_2 + 2CO_2 + H_2O \quad \text{A-7}$$

$$Fe^{2+} + \frac{1}{4}O_2 + 2HCO_3^- + \frac{1}{2}H_2O \xrightarrow{R_8} Fe(OH)_3 + 2CO_2 \quad \text{A-8}$$

$$2Fe^{2+} + MnO_2 + 2HCO_3^- + 2H_2O \xrightarrow{R_9} 2Fe(OH)_3 + Mn^{2+} + 2CO_2 \quad \text{A-9}$$

$$NH_4^+ + 2O_2 + 2HCO_3^- \xrightarrow{R_{10}} NO_3^- + 2CO_2 + 3H_2O \quad \text{A-10}$$

$$H_2S + 2O_2 + 2HCO_3^- \xrightarrow{R_{11}} SO_4^{2-} + 2CO_2 + 2H_2O \quad \text{A-11}$$

$$H_2S + 2CO_2 + MnO_2 \xrightarrow{R_{12}} Mn^{2+} + S^o + 2HCO_3^- \quad \text{A-12}$$

$$H_2S + 4CO_2 + 2Fe(OH)_3 \xrightarrow{R_{13}} 2Fe^{2+} + S^o + 4HCO_3^- + 2H_2O \quad \text{A-13}$$

$$FeS + 2O_2 \xrightarrow{R_{14}} Fe^{2+} + SO_4^{2-} \quad \text{A-14}$$

$$CH_4 + 2O_2 \xrightarrow{R_{15}} CO_2 + 2H_2O \quad \text{A-15}$$

$$CH_4 + CO_2 + SO_4^{2-} \xrightarrow{R_{16}} 2HCO_3^- + H_2S \quad \text{A-16}$$

$$Mn^{2+} + 2HCO_3^- \xrightarrow{R_{17}} MnCO_3 + CO_2 + H_2O \quad \text{A-17}$$

$$Fe^{2+} + 2HCO_3^- \xrightarrow{R_{18}} FeCO_3 + CO_2 + H_2O \quad \text{A-18}$$

$$Fe^{2+} + 2HCO_3^- + H_2S \xrightarrow{R_{19}} FeS + 2CO_2 + 2H_2O \quad \text{A-19}$$

$$CO_3^{2-} + \delta CO_2 + \delta H_2O + (1-\delta)H_2S \leftrightarrows (1+\delta)HCO_3^- + (1-\delta)HS^- \qquad 0 \le \delta \le 1 \quad \text{A-20}$$

solids (C in units of mass per unit volume solid sediment) and solutes (C in units of mass per unit volume porewater solution). The reaction rate R and the nonlocal transport term S in Equation 2 are expressed in units of mass constituent per unit time and per unit volume solid sediment or porewater. The parameter ζ is related to the sediment porosity, ϕ:

$\zeta = 1 - \phi$ for solids

$\zeta = \phi$ for solutes.

In Equation 2 the local transport fluxes have been separated in diffusive and advective fluxes. The diffusive processes in a sediment are molecular porewater diffusion, bulk sediment mixing, and hydrodynamic dispersion [4]. Bulk sediment mixing can result from the activity of surface dwelling and burrowing animals (bioturbation), or from the stirring action of waves and bottom currents.

Mass transfer over finite distances, during which the transported fluid or sediment is not substantially modified by small-scale mixing, is referred to as nonlocal transport. Nonlocal transport phenomena in sediments include bioirrigation (or flushing), deposit feeding, and wave pumping [5].

Analytical solutions to Equation 2 can be found for simple systems [4]. Some of the simplifying conditions include conservative behavior ($\Sigma\ R = 0$), steady state ($\partial\zeta C/\partial t = 0$) and the absence of nonlocal sediment reworking or irrigation ($\Sigma\ S = 0$). In general, however, we are interested in the time-dependent behavior of reactive, multicomponent sedimentary systems where the distributions of the various species are coupled to one another. Furthermore, nonlinear rate expressions may be required to describe certain of the microbial and geochemical processes. To solve coupled, nonlinear sets of conservation equations, numerical methods are necessary.

Standard finite-difference methods are intuitively the simplest way to numerically solve partial differential equations [6]. In one possible numerical scheme, the partial derivatives in Equation 2 are discretized as follows (only the concentration derivatives are shown):

$$\frac{\partial C}{\partial t} = \frac{C_i^{n+1} - C_i^n}{\Delta t} \tag{3}$$

$$\frac{\partial C}{\partial x} = \theta \left\{ \frac{C_i^{n+1} - C_{i-1}^{n+1}}{\Delta x} + (1-\alpha)\frac{C_{i+1}^{n+1} - C_{i-1}^{n+1}}{2\Delta x} \right\}$$

$$+ (1-\theta)\left\{ \frac{C_i^n - C_{i-1}^n}{\Delta x} + (1-\alpha)\frac{C_{i+1}^n - C_{i-1}^n}{2\Delta x} \right\} \qquad (4)$$

$$\frac{\partial^2 C}{\partial x^2} = \theta \left\{ \frac{C_{i+1}^{n+1} - 2C_i^{n+1} + C_{i-1}^{n+1}}{\Delta x^2} \right\} + (1-\theta)\left\{ \frac{C_{i+1}^n - 2C_i^n + C_{i-1}^n}{\Delta x^2} \right\} (5)$$

where Δx is the grid spacing in the vertical (depth) direction and Δt is the interval between time steps; C_i^n refers to the concentration of the constituent at the ith grid point and at time $n\Delta t$; α and θ are adjustable parameters with values between 0 and 1.

The parameters α and θ in Equations 4 and 5 allow to experiment with various discretization schemes. The case where $\alpha = 1$ is known as the *upstream-weighted* formulation, whereas $\alpha = 0$ gives the *central-difference* formulation. With $\theta = 0$, the method is completely *explicit*, while it is completely *implicit* for $\theta = 1$. By varying α and θ it is possible to play off accuracy (which is greater for the central difference approach) *versus* numerical stability (which is usually better for upstream-weighted methods). Equations 3-5 are written for a constant grid spacing, Δx. Finite-difference schemes, however, are easily amended to incorporate variable grid spacing. In sediments, the biogeochemical transformations are usually most intense in the top few centimeters. It may, therefore, be advantageous to compress the grid spacing in the top sediment and expand it at greater depths.

The choice of a numerical scheme depends ultimately on the physical nature of the problem at hand. For example, the Crank-Nicholson scheme, with $\theta = 1/2$, is well-suited for calculating distributions that are dominated by diffusional transport. For cases where advection is the principal transport mechanism, it may be necessary to combine a large value of α with a fine grid spacing to eliminate numerical oscillations in the vicinity of sharp concentration gradients. Numerical stability analysis can guide the development of appropriate finite-difference methods through a comparison of the magnitudes of transport and reaction parameters. The presence of non-linear reaction terms and the possibility of variable transport parameters, however, complicates such an analysis. Consider, for instance, the case of a stirred surface layer overlying unstirred sediment. (The stirring of the top layer may be induced by waves and bottom currents or by infaunal activity.) In this case, transport of solids may be dominated by random mixing in the top layer, and by sediment advection in the underlying sediment. Hence, a numerical scheme that works well for solids in the top sediment, may perform poorly at greater depths. It is therefore not surprising that designing finite difference schemes is often considered an art as much as a science.

The finite difference formulation of a transport-reaction equation of a porewater or solid sediment constituent can be transformed into a set of algebraic equations:

$$aC_{i-1}^{n+1} + bC_i^{n+1} + c_iC_{i+1}^{n+1} = d_i \qquad \text{with} \quad i = 1, 2, ..., N_x \qquad (6)$$

where N_x is the number of grid points. The coefficients a_i, b_i, c_i, d_i in Equation 6 are functions of the sedimentation rate, the dispersion-diffusion coefficient, the sediment porosity, the nonlocal transport source strengths, the reaction rates and the finite differences Δx and Δt. The solution of the system of Equations 6 is obtained through inversion of the coefficient matrix. Efficient and high accuracy solvers for linear systems of equations abound in the literature. In the present case, the choice of an inversion algorithm is dictated by the band form of the coefficient matrix, with all the non-zero elements located on the diagonal and two adjacent codiagonals.

The rate R in Equation 2 is positive when the constituent is produced during a biogeochemical reaction and negative when the constituent is consumed. Generally speaking, rate expressions depend on the concentrations of several reactive species present in the system. For example, the rates of oxidation of dissolved Mn^{2+} and Fe^{2+} ions diffusing upward into the oxidized surface layer of a sediment depend, among others, on the dissolved metal concentration and the level of dissolved oxygen (see section 4.3). Thus, within the framework of a general sediment model, the porewater and solid sediment distributions of all reactive species must be solved simultaneously.

4.   A MODEL FOR IRON AND MANGANESE IN SEDIMENTS

4.1   Reactive species

The first step in the construction of a model for Fe and Mn cycling in surface sediments is the choice of the reactive chemical species.

These may include the various dissolved, interfacial and solid forms of
the metals themselves, as well as porewater and solid sediment species
that affect the production or consumption of the metal species. Table 2
shows the dissolved and solid constituents used in the model. The list
was kept modest, while still allowing for a realistic representation of the
chemical complexity of the system.

Table 2. Dissolved and Solid Constituents. $MnCO_3$ and $FeCO_3$ may represent either the pure end-member minerals rhodocrosite and siderite, or components of a solid solution.	
Dissolved Constituents	O_2
	NO_3^-
	Mn^{2+}
	Fe^{2+}
	SO_4^{2-}
	NH_4^+
	CH_4
	HCO_3^-
	HS^-
Solid Constituents	$(CH_2O)_x(NH_3)_y(H_3PO_4)_z$
	MnO_2
	$Fe(OH)_3$
	$MnCO_3$
	$FeCO_3$
	FeS

The dissolved species are represented by their total concentrations.
Thus, Fe^{2+}, for example, stands for the sum of all dissolved ferrous
iron species. The transport-reaction equation of a dissolved constituent
written in terms of its total concentration, C_T, is,

$$\frac{\partial \phi C_T}{\partial t} = \frac{\partial}{\partial x}\left\{\sum_{i=1}^{N}\left(D_i \frac{\partial \alpha_i C_T}{\partial x}\right) - \omega \phi C_T\right\} - \phi \sum S_T + \phi \sum R_T \qquad (7)$$

where the subscript i refers to the N individual dissolved species (e.g., Fe^{2+}_{free}), $Fe(OH)^+$, Fe^{2+}-DOM, $Fe(HPO_4)^0$) that make up the total concentration, and α is the ratio of the concentration of the i-th species over the total concentration. The source and rate terms $\sum S_T$ and $\sum R_T$ describe the net production of the dissolved constituent. Thus, the homogeneous speciation reactions that transform one dissolved species of the constituent into another are not included in $\sum R_T$.

The phases $Fe(OH)_3$ and MnO_2 listed in Table 2 are idealized representations of the reactive, bioavailable fractions of solid Fe(III) and Mn(III, IV). Iron monosulfide, FeS, is the initial Fe(II) sulfide precipitate forming in anoxic sediments, as the result of sulfate reduction [7]. The kinetically-favored FeS is used in the model because it controls porewater iron concentrations in sulfide-rich surface sediments, rather than the thermodynamically more stable pyrite. Carbonate alkalinity production during anaerobic respiration may lead to the precipitation of carbonate minerals [2]. In freshwater sediments, relatively pure $FeCO_3$ and $MnCO_3$ may form [8]. In marine environments, the formation of solid solutions with $CaCO_3$ is more likely, especially for Mn(II) [9].

For the sake of simplicity, interfacial species of Fe and Mn are included with the solid species in the model. Adsorbed Fe and Mn species are important reaction intermediates in the redox transformations and mineral precipitation-dissolution reactions of Fe and Mn. In addition, they represent a critical component of the bioavailable fraction of the metals. Surface species can be added to the model by writing separate conservation equations for the total adsorbed concentrations of Fe and Mn. Further discussion on the inclusion of adsorption reactions in mathematical sediment models can be found elsewhere [4].

With the exception of particulate organic matter, the remaining species in Table 2 are porewater species. Taken together, the depth profiles of these species describe the vertical distribution of chemical conditions and metal speciation in the sediment.

4.2 Reactions

The decomposition of sedimentary organic matter consists of many enzymatic reactions involving a variety of organisms and a variety of intermediate compounds. Despite this complexity, it is possible to write overall degradation reactions which only take into account the initial reactants and the final products. The overall reactions considered in the model are listed in Table 1; they include oxic respiration (A-1), denitrification (A-2), manganese oxide reduction (A-3), iron (hydr)oxide reduction (A-4), sulfate reduction (A-5), and methane formation (A-6). The reactions are listed roughly in the sequence in which they occur with increasing depth in sediments. This succession reflects the order of decreasing energy yield of the oxidation reactions for sediment heterotrophs [4].

The degradation of sedimentary organic matter is often non-stoichiometric, i.e., the relative release rates of organically-bound elements during decomposition may differ from those predicted on the basis of the elemental ratios of the total particulate organic matter [2]. This reflects the fact that sedimentary organic matter is a complicated mixture of compounds that differ in their reactivities towards degradation. The addition of biomass produced *in-situ* by bacteria and macrofauna also modifies the composition and nature of sedimentary organic detritus. Thus, the stoichiometric x:y:z ratios in reactions A-1 to A-6 of Table 1 may vary with depth and time.

The oxidation of organic matter results in the production of reduced porewater species, Mn^{2+}, Fe^{2+}, NH_4^+, HS^-, CH_4, which migrate upward in the sediment and may reoxidize when encountering more oxidizing conditions (reactions A-7 to A-13, A15 and A16). Many of these reoxidation reactions are microbially-mediated (see section 4.3). In reactions A-12 and A-13 it is assumed that elemental sulfur is the sole sulfide oxidation product. The actual situation is more complicated: additional reaction products may include polysulfides, thiosulfate and sulfate.

Reactions A-17 to A-19 in Table 1 correspond to the precipitation of carbonate and sulfide minerals. Burial of these mineral phases represent permanent sinks for reactive Fe and Mn in the sediment. Reaction A-14 accounts for the oxidation of authigenically formed FeS that is brought back up into the aerobic zone by sediment mixing.

The principal weak acids in sediments are carbonic acid and, when sulfate reduction is important, hydrogen sulfide. It is assumed in the model that the pH of porewaters is determined by the extent of dissociation of these two acids. Porewater buffering is incorporated in reactions A-1 to A-19 by assuming that protons produced or consumed in the reactions are balanced by H^+ uptake or release during rapid bicarbonate-carbonic acid inter-conversion. The reaction stoichiometries A-1 to A-19, therefore, describe the effect of the irreversible reactions on porewater alkalinity. Equilibrium reaction A-20 describes local pH balance and alkalinity conservation.

4.3   Rate laws

With the exception of reaction A-20, all reactions in Table 1 are irreversible processes and, hence, rate expressions are needed to determine their effect on porewater and solid sediment chemistry. Table 3 compiles the rate laws used in the model.

Most kinetic models for organic matter degradation in sediments derive from the simple first-order kinetic model introduced by Berner [4]:

$$R_c = -\frac{d[CH_2O]}{dt} = k_c[CH_2O]_m \tag{8}$$

where R_c is the net rate of organic carbon oxidation, k_0 is a first-order rate coefficient, $[CH_2O]$ is the concentration of organic matter, and the subscript m refers to the metabolizable fraction of the organic matter. Equation 8 states that the principal controls on the rate of oxidation of organic matter by benthic metabolism are the amount and the reactivity (or bioavailability) of the organic matter itself.

In the simplest possible case, the rate coefficient k_c in Equation 8 is assigned a single value, representative of the "average" reactivity of the sedimentary organic matter over the depth range of interest. This is the approach adopted here. More sophisticated models have been proposed which take into account the variability of organic matter reactivity with advancing degradation [2,3,10,11].

Equation 8 calculates the total rate of organic carbon oxidation. This rate can be decomposed in the contributions of the individual metabolic pathways represented by reactions A-1 to A-6 in Table 1. To this end, the fraction, f_i, of the i-th metabolic pathway is introduced:

$$f_i = \frac{R_i}{R_c} \tag{9}$$

where R_i is the rate of carbon oxidation by reaction A-i, i = 1 to 6. At any depth and time, the following condition applies:

$$\sum_{i=1}^{6} f_i = 1 \tag{10}$$

Table 3.   Rate Laws Used in the Model; Square Brackets Denote Concentrations. See text for discussion.

$$R_i = \frac{f_i}{x} k_c [CH_2O]_m \qquad\qquad i = 1, ..., 6 \qquad\qquad\qquad \text{B-1 to B-6}$$

$$R_7 = k_7 [Mn^{2+}][O_2] \qquad\qquad\qquad\qquad\qquad\qquad \text{B-7}$$

$$R_8 = k_8 [Fe^{2+}][O_2] \qquad\qquad\qquad\qquad\qquad\qquad \text{B-8}$$

$$R_9 = k_9 [Fe^{2+}][MnO_2] \qquad\qquad\qquad\qquad\qquad\qquad \text{B-9}$$

$$R_{10} = k_{10} [NH_4^+][O_2] \qquad\qquad\qquad\qquad\qquad\qquad \text{B-10}$$

$$R_{11} = k_{11} [O_2] TS \qquad\qquad TS = [HS^-] + [H_2S] \qquad\qquad \text{B-11}$$

$$R_{12} = k_{12} [MnO_2] TS \qquad\qquad\qquad\qquad\qquad\qquad \text{B-12}$$

$$R_{13} = k_{13} [Fe(OH)_3] TS \qquad\qquad\qquad\qquad\qquad\qquad \text{B-13}$$

$$R_{14} = k_{14} [FeS][O_2] \qquad\qquad\qquad\qquad\qquad\qquad \text{B-14}$$

$$R_{15} = k_{15} [CH_4][O_2] \qquad\qquad\qquad\qquad\qquad\qquad \text{B-15}$$

$$R_{16} = k_{16} [CH_4][SO_4^{2-}] \qquad\qquad\qquad\qquad\qquad\qquad \text{B-16}$$

$$R_{17} = k_{17}\delta_{17}\left(\frac{[Mn^{2+}][CO_3^{2-}]}{K'_{s,MnCO_3}} - 1 \right) \qquad\qquad\qquad \text{B-17}$$

$$R_{18} = k_{18}\delta_{18}\left(\frac{[Fe^{2+}][CO_3^{2-}]}{K'_{s,FeCO_3}} - 1 \right) \qquad \begin{array}{l} \delta = 0 \ \text{for} \ IAP \leq K'_s \\ \delta = 1 \ \text{for} \ IAP > K'_s \end{array} \qquad \text{B-18}$$

$$R_{19} = k_{19}\delta_{19}\left(\frac{[Fe^{2+}][HS^-]}{a_{H^+} K'_{s,FeS}} - 1 \right) \qquad\qquad\qquad \text{B-19}$$

To calculate the fraction f_i of a given respiratory pathway (i = 1 to 5), the model first determines whether that pathway is repressed by energetically more favorable pathways. If not, the model checks whether the rate of the i-th pathway is limited or not by the availability of the i-th external electron acceptor. The decision algorithm is based on a modified Monod or Michaelis-Menten formulation [12], which assumes that for each of the respiratory pathways there exists a critical or limiting concentration of the external oxidant. When the concentration of the oxidant exceeds the critical value, the rate of the metabolic pathway is independent of the concentration of the oxidant. Furthermore, the energetically less favorable pathways are inhibited. When the oxidant concentration drops below the critical level, the rate of the respiratory pathway becomes limited by the availability of the terminal electron acceptor, and energetically less powerful oxidants start to be utilized by the benthic heterotrophs. The computational scheme allows for a smooth transition from a zone in which a particular degradation pathway dominates another.

Application of the above approach requires a set of limiting concentrations for the successive external electron acceptors: O_2, NO_3^-, Mn(VI), Fe(III), SO_4^{2-}. Table 4 lists the ranges of the limiting concentrations reported for natural aquatic environments. In those cases where no values could be found in the literature, the ranges given in Table 4 produce depth distributions of reactive species that are comparable to measured profiles. For instance, the high limiting concentration of microbially reducible Fe(III) is consistent with the reported persistence of amorphous Fe(III) in highly reducing anoxic sediments [8,19].

The abiotic oxidation kinetics of dissolved Fe^{2+} and Mn^{2+} have been studied extensively [20,21]. These studies suggest rate laws B-7 and B-8 for the rate of oxidation of the divalent cations by oxygen. The rate coefficients k_7 and k_8 are dependent on pH and on the availability of binding sites for dissolved cations. In natural aquatic environments, oxidation of Mn^{2+} and Fe^{2+} may be microbially-mediated [22]. The limited data available suggest that the kinetics of the microbial pathways are sensitive to the same environmental parameters as the abiotic oxidation reactions. For instance, at low reactant concentrations, the microbial oxidation of Mn^{2+} correlates positively with the concentrations of dissolved Mn^{2+} and oxygen [23,24]. Equations B-7 and B-8 are, therefore, assumed to describe the rates of oxidation of Mn^{2+} and Fe^{2+} by oxygen in sediments, whether the reactions are chemical, microbiological or some combination of both. A similar bimolecular rate law is assumed for the oxidation of ferrous iron by manganese oxides.

Table 4. Limiting Concentrations of Terminal Electron Acceptors in Respiration Pathways.

Pathway	Limiting reactant	Limiting concentration	Environment	Ref.
aerobic respiration	dissolved O_2	1-10 μM	marine	12, 13
			freshwater	
denitrification	dissolved NO_3	50-80 μM	marine	14, 15
		4-20 μM	freshwater	2, 13, 16
Mn reduction	solid Mn(IV)	1-10 μmol/g	marine	
			freshwater	
Fe reduction	solid Fe(III)	1-50 μmol/g	marine	
			freshwater	
sulfate reduction	dissolved SO_4	1600 μM	marine	17
		60-300 μM	freshwater	18

Studies of dissolved sulfide oxidation by oxygen suggest the empirical rate expression B-11 [25]. Existing data indicates that the oxidation rate of sulfide by Fe or Mn oxides correlates positively with the sulfide concentration and with the oxide surface areas [26,27]. Similarly, the oxidation rate of solid Fe(II) sulfides in the presence of oxygen depends on the concentration of the oxidant and the availability of mineral surface sites [28]. Because surface reactions are typically the rate determining steps in oxidative and reductive dissolution processes, the empirical rate laws may exhibit complex, non-linear dependencies on the dissolved oxidant or reductant concentration. However, for the sake of simplicity, non-unity reaction orders are not included in rate expressions B-12, B-13 and B-14.

In the absence of detailed kinetic information, the rates of nitrification (A-10) and methane oxidation (A-15 and A-16) are calculated using simple bimolecular rate Equations (B-10, B-15 and B-16).

Linear rate laws are used for the precipitation of carbonate and sulfide phases from supersaturated porewaters (B-17, B-18 and B-19). Strictly speaking, neither the rate coefficients nor the apparent solubility constants in the rate laws are true constants. These parameters depend on the compositions of porewaters and solids. In addition, the apparent rate constants vary with changes in the reactive mineral surface areas. These effects are ignored here.

4.4 Transport

Molecular diffusion coefficients of dissolved species in sediments must be corrected for sediment tortuosity, temperature, and porewater viscosity [4,29]. Temperature coefficients for the species considered in this model are listed in Table 5. For recent, unconsolidated muds the following tortuosity correction has been proposed [31]:

$$D_{sed} = \phi^2 D_{sol} \tag{11}$$

where ϕ is the porosity, and D_{sol} and D_{sed} are the diffusion coefficients in solution and in the sediment, respectively.

Sediment reworking is frequently modeled as a random mixing process, quantified by a mixing coefficient, D_{mix}. It affects both the solid sediment and the associated porewaters. When caused by macrofauna, the process is called bioturbation. It has been shown that

Table 5. Molecular Diffusion Coefficients in Water at Infinite Dilution. The coefficients α describe the temperature dependence of the diffusion coefficients according to the formula:

$$D_0(t°C) = D_0(0°C)\{1+\alpha\ t\}$$

where t is temperature in °C. The lower part of the table gives the viscosity-based correction for calculating diffusion coefficients in seawater (35 S‰) [Ref. 29,30].

Species	D_0(0°C) in cm²/a	α (°C⁻¹)
O_2	296	0.060
NO_3^-	307	0.038
Mn^{2+}	96	0.050
Fe^{2+}	106	0.044
SO_4^{2-}	156	0.045
NH_4^+	308	0.041
CH_4	235	0.052
$CO_{2,aq}$	249	0.060
HCO_3^-	169	0.048
CO_3^{2-}	137	0.047
H_2S	258	0.060
HS^-	305	0.031

	Temperature (°C)			
	0	10	20	30
$D_{seawater}/D_o$	0.95	0.94	0.93	0.92

in oxygenated marine depositional environments the mixing coefficient correlates positively with the sediment accumulation rate [2,42]. The reason is that a higher sedimentation rate generally corresponds to a higher deposition flux of organic matter, i.e., a larger food supply for the benthic macrofauna and, thus, a higher intensity of bioturbation. The following equation predicts macrofaunal mixing coefficients in marine environments within an order of magnitude:

$$\log D_{mix} = 1.63 + 0.851 \log \omega \tag{12}$$

where D_{mix} is expressed in cm^2/a and ω in cm/a. The use of Equation 12 is restricted to sediments accumulating in oxic bottom waters. Benthic macrofaunal activity slows down when oxygen levels in the bottom waters drop below 30% saturation; it essentially ceases at oxygen concentrations below 0.2 ml/L [32]. Additionally, Equation 12 does not apply to areas, such as deltas and certain nearshore environments, where an extremely rapid accumulation of detrital material disrupts the establishment of a stable benthic animal population (typically, $\omega > 5$ cm/a).

The total diffusion-dispersion coefficient appearing in the transport-reaction Equation 2 includes the contributions of all random mixing processes. Thus, in a stirred sediment we have:

$$D = D_{sed} + D_{mix} \qquad \text{for a dissolved constituent}$$

$$D = D_{mix} \qquad \text{for a solid constituent}$$

In organic-rich, heavily bioturbated sediments, D_{mix} may be of the same order of magnitude as, or even exceed the molecular diffusion coefficient, D_{sed}, of dissolved species.

Changes in speciation of a dissolved constituent affect its overall diffusional flux, as can be inferred from Equation 7. If a single species dominates the speciation of a dissolved constituent over the depth range of interest, Equation 7 can be simplified to:

$$\frac{\partial \phi C_T}{\partial t} = \frac{\partial}{\partial x} \left\{ \phi \left(\tilde{D}_i \frac{\partial C_T}{\partial x} \right) - \omega \phi C_T \right\} + \phi \sum S_T + \phi \sum R_T \tag{13}$$

where $\tilde{D}$ is the dispersion-diffusion coefficient of the dominant species. When several species are important, $\tilde{D}$ is sometimes replaced in Equation 13 by the weighted dispersion-diffusion coefficient:

$$\overline{D} = \sum_{i=1}^{N} \alpha_i D_i \tag{14}$$

This approach, however, is strictly valid only when the speciation of the constituent does not vary with depth ($\partial \alpha_i / \partial x = 0$). Although Equation 13, with $\tilde{D}$ or $\overline{D}$, is widely used by sediment modelers it must be remembered that it represents a special case of the more general transport-reaction Equation 7.

Irrigation by tube dwelling animals or wave action can cause a several-fold enhancement of the solute exchanges at the water-sediment interface [33]. A number of models for sediment irrigation have been presented in the literature [34]. The simplest nonlocal model is

$$S(x) = \alpha_x (C_o - C_x) \tag{15}$$

where C_0 and C_x are the concentrations of the constituent in the overlying water and at depth x in the sediment, α_x is the solute exchange coefficient at depth x and $S(x)$ is the irrigation source strength. Values between 2 and 300 a^{-1} have been reported for exchange coefficients in nearshore sediments [35]. Some of the highest values may in fact correspond to wave-induced irrigation rather than bio-irrigation.

4.5   Transport-reaction equations

For each constituent listed in Table 2 a transport-reaction equation is solved. Table 6 lists the equations used in the model. The constituents are coupled to one another through the reaction terms. The exception is organic carbon whose distribution in the sediment is obtained independently from any other species in the system. The equations for total dissolved carbonate (TC) and total dissolved H$_2$S (TS) are discussed in the next section. The goal of the simulations is primarily to investigate effects caused by chemical coupling. The set of equations is, therefore, simplified by assuming steady state and by treating sediment porosity and density as constants.

Table 6. Transport-reaction Equations. Assumptions include steady state, plus constant values for porosity, sediment accumulation rate (ω), sediment mixing coefficient (d_{mix}) and solute exchange coefficient (α_{irrig}). Symbols: ∂_t, ∂_x: time and depth partial derivatives; x_{mix}: depth of the mixed surface layer. Concentrations are expressed in mols per unit volume total sediment.

$\partial_t[CH_2O]_{lm} = 0 = D_{solids}\partial_x^2[CH_2O]_{lm} - \omega\partial_x[CH_2O]_{lm} - R_C$ (C-1)

$\partial_t[MnO_2] = 0 = D_{solids}\partial_x^2[MnO_2] - \omega\partial_x[MnO_2] - 2xR_3 + R_7 - R_9 - R_{12}$ (C-2)

$\partial_t[Fe(OH)_3] = 0 = D_{solids}\partial_x^2[Fe(OH)_3] - \omega\partial_x[Fe(OH)_3] - 4xR_4 + R_8 + 2R_9 - 2R_3$ (C-3)

$\partial_t[MnCO_3] = 0 = D_{solids}\partial_x^2[MnCO_3] - \omega\partial_x[MnCO_3] + R_{17}$ (C-4)

$\partial_t[FeCO_3] = 0 = D_{solids}\partial_x^2[FeCO_3] - \omega\partial_x[FeCO_3] + R_{18}$ (C-5)

$\partial_t[FeS] = 0 = D_{solids}\partial_x^2[FeS] - \omega\partial_x[FeS] - R_{14} + R_{19}$ (C-6)

$\partial_t[O_2] = 0 = D_{O_2}\partial_x^2[O_2] - \omega\partial_x[O_2] + \alpha_x([O_2]_0 - [O_2]) - (x+2y)R_1 - \frac{1}{2}R_7 - \frac{1}{4}R_8 - 2(R_{10} + R_{11} + R_{14} + R_{15})$ (C-7)

$\partial_t[NO_3^-] = 0 = D_{NO_3^-}\partial_x^2[NO_3^-] - \omega\partial_x[NO_3^-] + \alpha_x([NO_3^-]_0 - [NO_3^-]) + yR_1 - \frac{(4x+3y)}{5}R_2 + R_{10}$ (C-8)

$\partial_t[Mn^{2+}] = 0 = D_{Mn^{2+}}\partial_x^2[Mn^{2+}] - \omega\partial_x[Mn^{2+}] + \alpha_x([Mn^{2+}] - [Mn^{2+}]_0) + 2xR_3 - R_7 + R_9 + R_{12} - R_{17}$ (C-9)

$\partial_t[Fe^{2+}] = 0 = D_{Fe^{2+}}\partial_x^2[Fe^{2+}] - \omega\partial_x[Fe^{2+}] + \alpha_x([Fe^{2+}] - [Fe^{2+}]_0) - 4xR_4 - R_8 - 2R_9 + 2R_{13} + R_{14} - R_{18} - R_{19}$ (C-10)

$\partial_t[SO_4^{2-}] = 0 = D_{SO_4^{2-}}\partial_x^2[SO_4^{2-}] - \omega\partial_x[SO_4^{2-}] + \alpha_x([SO_4^{2-}]_0 - [SO_4^{2-}]) - \frac{x}{2}R_5 + R_{11} + R_{14} - R_{16}$ (C-11)

$\partial_t[NH_4^+] = 0 = D_{NH_4^+}\partial_x^2[NH_4^+] - \omega\partial_x[NH_4^+] + \alpha_x([NH_4^+] - [NH_4^+]_0) - y(R_3 + R_4 + R_5 + R_6) - R_{10}$ (C-12)

$\partial_t[CH_4] = 0 = D_{CH_4}\partial_x^2[CH_4] - \omega\partial_x[CH_4] + \alpha_x([CH_4]_0 - [CH_4]) + \frac{x}{2}R_6 - R_{15} - R_{16}$ (C-13)

$x \leq x_{mix}$: $D_{solids} = D_{mix}$ $D_i = D_{sec,i} - D_{mix}$ $\alpha_x = \alpha_{irrig}$

$x > x_{mix}$: $D_{solids} = 0$ $D_i = D_{sec,i}$ $\alpha_x = 0$

4.6   Alkalinity and pH

It is assumed in the model that the pH in surface sediments is buffered by the carbonate and sulfide acid-base systems. It is, therefore, useful to cast the acid-base chemistry of the porewaters in terms of the following conservative parameters:

$$TC = \left[CO_2^*\right] + \left[HCO_3^-\right] + \left[CO_3^{2-}\right] \tag{16}$$

$$TS = [H_2S] + \left[HS^-\right] \tag{17}$$

$$ALK = \left[HCO_3^-\right] + 2\left[CO_3^{2-}\right] + \left[HS^-\right] + \left[OH^-\right] - \left[H^+\right] \tag{18}$$

where TC stands for total dissolved CO_2 (CO_2^* is the sum of hydrated and unhydrated dissolved CO_2), TS for total dissolved H_2S and ALK for alkalinity. The depth distributions of the three parameters obey Equations C-14, C-15 and C-16 in Table 7. The rate expressions in the transport-reaction equations describe the production of TC, TS and ALK by the irreversible reactions A-1 to A-19.

To determine the porewater profiles of TC, TS, ALK and pH, the transport-reaction Equations C-14 to C-16 are coupled to the condition of local acid-base equilibrium (see Tables 7 and 8). Starting from an initial guess of the pH profile, Equations D-1 to D-5 are combined with C-14 and C-15 to calculate profiles of TC and TS. With this information, Equations C-16, D-6, D-7 and D-8 can be solved for the alkalinity distribution. An improved pH profile can now be derived from the equilibrium expression:

$$\left\{H^+\right\} =$$

$$\frac{K_1'}{2ALK_c} \left\{ (TC - ALK_c) + \left[(TC - ALK_c)^2 - 4ALK_c \frac{K_2'}{K_1'}(ALK_c - 2TC) \right]^{\frac{1}{2}} \right\} \tag{19}$$

Table 7. Pore Water Acid-base Chemistry is Calculated by Coupling the Irreversible Production and Consumption of Total Dissolved Carbonate (TC), Total Dissolved Sulfide (TS) and Alkalinity (ALK) to Local Acid-base Equilibrium. Equations D-1 to D-8 express the equilibrium species distribution in the dissolved carbonate-sulfide system. For further discussion on the dissociation constants, see Table 8.

$$\partial_t(TC) = 0 = D_{CO_2}\partial_x^2(\alpha_0 TC) + D_{HCO_3^-}\partial_x^2(\alpha_1 TC) + D_{CO_3^{2-}}\partial_x^2(\alpha_2 TC) - \omega\partial_x(TC) + \alpha_x(TC_0 - TC)$$
$$+ x(R_1 + R_2 + R_3 + R_4 + R_5) + \tfrac{y}{2}R_6 + R_{15} + R_{16} - F_{17} - R_{18} \qquad \text{C-14}$$

$$\partial_t(TS) = 0 = D_{H_2S}\bar{c}_x^2(\alpha_{0,S}TS) + D_{HS^-}\partial_x^2(\alpha_{1,S}TS) - \omega\bar{c}_x(TS) + c_x(TS_0 - TS) + \tfrac{y}{2}R_5 - R_{11} - R_{12} - R_{13} + R_{16} - R_{19} \qquad \text{C-15}$$

$$\partial_t(ALK) = 0 = D_{HCO_3^-}\partial_x^2(\beta_1 ALK) + 2D_{CO_3^{2-}}\partial_x^2(\beta_2 ALK) + D_{HS^-}\partial_x^2(\beta_{1,S}ALK) - \omega\partial_x(ALK) + \alpha_x(ALK_0 - ALK)$$
$$- (x + 2z)R_1 + \tfrac{(-z+3y-10z)}{z}\tfrac{y}{z}R_2 + (4x + y - 2z)R_3 + (8x + y - 2z)R_4 + (x + y - 2z)R_5 + (y - 2z)R_6$$
$$- 2(R_7 + R_8 + R_9 + R_{10} + R_{11} + R_{17} + R_{18} - R_{19}) + \sum R_{12} + 2R_{13} + R_{16}) \qquad \text{C-16}$$

$$\alpha_0 = \frac{[CO_2^*]}{TC} = \left(1 + \frac{K_1'}{\{H^+\}} + \frac{K_1'K_2'}{\{H^+\}^2}\right)^{-1} \qquad \text{D-1}$$

$$\beta_1 = \frac{[HCO_3^-]}{ALK} = \frac{\alpha_1 TC}{(\alpha_1 + 2\alpha_2)TC + \alpha_{1,S}TS} \qquad \text{D-6}$$

$$\alpha_1 = \frac{[HCO_3^-]}{TC} = \left(\frac{\{H^+\}}{K_1'} + 1 + \frac{K_2'}{\{H^+\}}\right)^{-1} \qquad \text{D-2}$$

$$\beta_1 = \frac{[CO_3^{2-}]}{ALK} = \frac{\alpha_2 TC}{(\alpha_1 + 2x_2)TC + \alpha_{1,S}TS} \qquad \text{D-7}$$

$$\alpha_2 = \frac{[CO_3^{2-}]}{TC} = \left(\frac{\{H^+\}^2}{K_1'K_2'} + \frac{\{H^+\}}{K_2'} + 1\right)^{-1} \qquad \text{D-3}$$

$$\beta_{1,S} = \frac{[HS^-]}{ALK} = \frac{\alpha_{1,S}TS}{(\alpha_1 + 2x_2)TC + \alpha_{1,S}TS} \qquad \text{D-8}$$

$$\alpha_{0,S} = \frac{[H_2S]}{TS} = \left(1 + \frac{K_{1,S}'}{\{H^+\}}\right)^{-1} \qquad \text{D-4}$$

$$\{H^+\} : H^- \text{activity}$$

$$\alpha_{1,S} = \frac{[HS^-]}{TS} = \left(1 + \frac{\{H^+\}}{K_{1,S}'}\right)^{-1} \qquad \text{D-5}$$

$$[CO_2^*] = [CO_{2,aq}] + [H_2CO_3]$$

Table 8. Dissociation Constants of Carbonic Acid and Hydrogen Sulfide at 1 atm. In freshwater environments, the temperature-corrected values of the thermodynamic constants (E-4 to E-6) are combined with calculated total activity coefficients. In marine environments, apparent dissociation constants are given directly by the empirical formulas E-7 to E-9. The square brackets denote total concentrations (free plus complexed species). Symbols: γ: total activity coefficient; I: ionic strength; S: salinity; Cl: chlorinity; T: absolute temperature; log = $\log_{10}$. [Ref. 36-40].

$$K'_1 = \frac{\{H^+\}[HCO_3^-]}{[CO_2^*]} = K_1 \frac{\gamma_{CO_2^*}}{\gamma_{HCO_3^-}} \quad (E-1); \qquad K'_2 = \frac{\{H^+\}[CO_3^{2-}]}{[HCO_3^-]} = K_2 \frac{\gamma_{HCO_3^-}}{\gamma_{CO_3^{2-}}} \quad (E-2); \qquad K'_{1,S} = \frac{\{H^+\}[HS^-]}{[H_2S]} = K_{1,S} \frac{\gamma_{H_2S}}{\gamma_{HS^-}} \quad (E-3)$$

Thermodynamic constants:

$$-\log K_1 = -126.3405 + 6320.81/T + 45.057 \log T \qquad \text{E-4}$$

$$-\log K_2 = -90.1833 + 5143.69/T + 33.648 \log T \qquad \text{E-5}$$

$$-\log K_{1,S} = 32.55 + 1519.44/T - 15.672 \log T + 0.02722T \qquad \text{E-6}$$

Apparent constants in seawater:

$$-\log K'_1 = -13.7201 + 0.031334T + 3235.67/T + 1.300 \times 10^{-5}(ST) - 0.1032\sqrt{S} \qquad \text{E-7}$$

$$-\log K'_2 = 5371.9645 + 1.671221T + 0.22913S + 18.3802 \log(S) - 128375.28/T - 2194.3055 \log(T)$$
$$\qquad\qquad -8.0944 \times 10^{-4}(ST) - 5617.11 \log(S)/T + 2.136(S/T) \qquad \text{E-8}$$

$$-\log K'_{1,S} = 2.527 + 1359.96/T - 0.206S^{1/3} \qquad \text{E-9}$$

$$S(\%_{oo}) = 1.80655 Cl(\%_{oo}) \qquad\qquad I = 0.00147 + 0.03590 Cl + 0.000068 Cl^2$$

where ALK_c stands for the carbonate alkalinity, defined as:

$$ALK_c = \left[HCO_3^-\right] + 2\left[CO_3^{2-}\right] \tag{20}$$

The procedure is iterated until convergence of all profiles.

4.7   Boundary conditions

For all solutes the upper boundary condition ($x = 0$) is the bottom water concentration. For solid sediment constituents, the deposition flux from the water column is specified. The flux continuity condition for a solid at the water-sediment interface is

$$F_0 = -D\left.\frac{\partial \hat{C}}{\partial x}\right|_{x=0} + \omega\hat{C}_0 \tag{21}$$

where F_0 is the deposition flux (mass per unit surface area sediment per unit time), D is the sediment reworking coefficient and $\hat{C}_0$ is the concentration of the constituent at the interface (mass per unit volume total sediment). This boundary condition is easily included in the finite-difference scheme discussed previously by using a downstream-weighted formulation for the depth derivative of $\hat{C}$ at the water-sediment interface.

The numerical calculations are performed down to a sufficiently great depth in the sediment where it may be assumed that reactions have essentially ceased.   The concentration gradients of the various constituents are then set equal to zero at the lower boundary. When a sediment consists of a stirred top layer overlying a non-stirred deeper layer, additional boundary conditions are needed to insure flux continuity at the boundary separating the layers.

5.   SIMULATIONS

5.1   Validation of the numerical model

A complete error analysis is beyond the scope of this paper. Nonetheless, in order to illustrate the performance of the finite difference method, the results of numerical calculations are compared to an analytical solution. The case considered is the steady state depth distribution of organic carbon, where a single rate constant, k_c, characterizes the decomposition kinetics. The governing equation is C-1. The following set of boundary conditions applies:

$$x = 0 \qquad F_m = -D\frac{\partial[CH_2O]_m}{\partial x}\bigg|_{x=0} + \omega[CH_2O]_{m,x=0}$$

$$x = x_{mix} \qquad D\frac{\partial[CH_2O]_m}{\partial x}\bigg|_{x=x_{mix}} = 0 \qquad\qquad (22)$$

$$x \to \infty \qquad \frac{\partial[CH_2O]_m}{\partial x}\bigg|_{x=\infty} = 0$$

where F_m is the deposition flux of metabolizable organic carbon and x_{mix} is the depth of the stirred surface layer. The analytical solutions for the organic carbon distribution are:

$$0 \leq x \leq x_{mix} \qquad \partial[CH_2O]_m = K^{-1}F_m\left\{\beta e^{\alpha(x-x_{mix})} - \alpha e^{\beta(x-x_{mix})}\right\} \quad (23)$$

$$x > x_{mix} \qquad \partial[CH_2O]_m = K^{-1}F_m(\beta - \alpha)e^{\gamma(x-x_{mix})} \qquad (24)$$

where:

$$\alpha = \frac{\omega - \sqrt{\omega^2 + 4k_cD_{mix}}}{2D_{mix}} \qquad\qquad (25)$$

$$\beta = \frac{\omega + \sqrt{\omega^2 + 4k_cD_{mix}}}{2D_{mix}} \qquad\qquad (26)$$

$$\gamma = \frac{k_c}{\omega} \tag{27}$$

$$K = (\beta D_{mix} - \omega)\alpha e^{-\beta x_{mix}} - (\alpha D_{mix} - \omega)\beta e^{-\alpha x_{mix}} \tag{28}$$

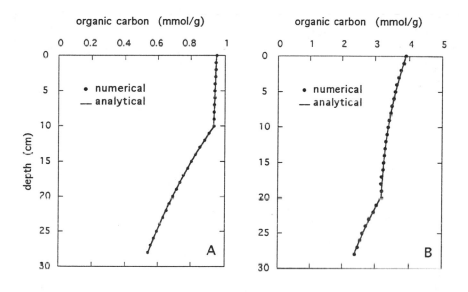

Figure 2.   *Comparison of Depth Profiles of Organic Carbon Concentration Calculated with an Analytical Solution and with the Numerical, Finite-difference Method (see text). Two cases are considered: a slowly accumulating pelagic sediment (A) and a rapidly deposited, organic-rich sediment (B). Conditions: sedimentation rate (cm/a): 0.005 (A), 1.0 (B); organic carbon deposition flux (mol/cm²/a): 3×10⁻⁶ (A), 3×10⁻³ (B); deposited organic carbon that is metabolizable (%): 90 (A), 75 (B); rate constant of decomposition (a⁻¹): 1.5×10⁻⁴ (A), 0.1 (B); particle mixing coefficient (cm²/a): 0.4 (A), 40 (B); depth of mixed layer (cm): 10 (A), 20 (B); porosity (%): 80 (A & B); dry density (g/cm³): 2.5 (A & B).*

Table 9. Environmental, Sedimentological, Biogeochemical, and Kinetic Parameters for Model Calculations of Representative Sediments. The values used were obtained from existing field-based data (for a compilation see [42]) or otherwise constrained using experimental data. Direct determinations for most of the rate constants in the table do not exist. The values reported here provide reasonable agreement between calculated and observed profiles of pore water and solid sediment properties. Much more work is needed, however, to reduce uncertainties on the numerical values of rate constants of biogeochemical processes.

	Deep-Sea	Shelf	Coastal/ Estuarine	Oligotrophic lake	Eutrophic lake
t (°C)	2	10	18	13	13
S (‰)	35	35	20	0.005	0.007
ω (cm/a)	0.001	0.04	1	0.1	0.4
ϕ (%)	80	85	90	85	85
ρ (g/cm^3)	2.5	2.5	2.4	2.5	2.4
D_{mix} (cm^2/a)	0.1	3	10	1.0	0.01
x_{mix} (cm)	20	30	30	10	10
α (a^{-1})	0	5	10	10	0
$[O_2]_0$ (μM)	180	250	275	180	30
$[NO_3]_0$ (μM)	30	15	10	15	60
$[Mn^{2+}]_0$ (μM)	0	0	0	0	40
$[Fe^{2+}]_0$ (μM)	0	0	0	0	0.5
$[SO_4]_0$ (mM)	28	28	16	0.2	0.2
$[NH_4]_0$ (μM)	0	0	0	0	5
pH_0	8.1	8.1	7.9.	7.9	7.5
Alk_0 (meq/dm^3)	2.4	2.4	5.0	1.0	5.0
TS_0 (μM)	0	0	0	0	0.2
F_c (μmol/cm^2/a)	10	100	1500	50	1000
% metabol. carbon	70	80	70	75	75
k_c (a^{-1})	0.003	0.01	0.1	0.1	0.1
x	200	200	106	150	400
y	21	21	11	16	16
z	1	1	1	1	1
$K_m^{O_2}$ (μM)	8	8	8	8	8
$K_m^{NO_3^-}$ (μM)	10	10	1	10	10
$K_m^{Mn^{IV}}$ (μmol/g)	2	2	2	2	2
$K_m^{Fe^{III}}$ (μmol/g)	5	5	5	5	5
$K_m^{SO_4^{2-}}$ (mM)	1	1	1	1	1
F_{Mn}^{IV} (μmol/cm^2/a)	4×10^{-5}	0.012	4	0.01	4
F_{Fe}^{III} (μmol/cm^2/a)	1×10^{-5}	0.1	10	0.02	7
k_7 (M^{-1} a^{-1})	1×10^9	2×10^9	2×10^9	2×10^9	2×10^9
k_8 (M^{-1} a^{-1})	1×10^9	2×10^9	2×10^9	2×10^9	2×10^9
k_9 (M^{-1} a^{-1})	1×10^4	2×10^8	2×10^8	1×10^6	1×10^6
k_{10} (M^{-1} a^{-1})	1×10^7	1.5×10^7	1.5×10^7	2×10^7	2×10^7
k_{11} (M^{-1} a^{-1})	3×10^8	6×10^8	6×10^8	6×10^8	6×10^8
k_{12} (M^{-1} a^{-1})	1×10^4	1×10^4	1×10^4	1×10^4	1×10^4
k_{13} (M^{-1} a^{-1})	1×10^4	1×10^4	1×10^4	1×10^4	1×10^4
k_{14} (M^{-1} a^{-1})	2×10^7	2.2×10^7	2.2×10^7	2×10^7	2×10^7
k_{15} (M^{-1} a^{-1})	1×10^{10}	1×10^{10}	1×10^{10}	1×10^{10}	1×10^{10}
k_{16} (M^{-1} a^{-1})	1×10^{10}	1×10^{10}	1×10^{10}	1×10^{10}	1×10^{10}
k_{17} (M a^{-1})	0	1×10^{-6}	1×10^{-6}	1×10^{-6}	1×10^{-6}
k_{18} (M a^{-1})	0	1×10^{-6}	1×10^{-6}	1×10^{-6}	1×10^{-6}
k_{19} (M a^{-1})	0	5×10^{-6}	5×10^{-6}	1×10^{-6}	1×10^{-6}

The numerical solutions shown in Figure 2 were obtained using an upstream-weighted formulation for the first derivatives. An inspection of the figure shows that the finite difference calculations are virtually indistinguishable from the analytical profiles. Thus, the numerical model provides an accurate description of the organic carbon concentrations in the mixing-dominated regime of the surface layer, as well as in the advection-dominated regime of the deeper sediment. Mass is conserved across the boundary between the mixed layer and the unstirred sediment. This is an important result since an ill-designed finite difference scheme will typically betray itself in the vicinity of abrupt changes in forcing parameters.

5.2   Representative sediments

The distribution and cycling of iron, manganese and associated reactive species are simulated in a number of "representative" depositional environments. Sediment properties, bottom water conditions, deposition fluxes, transport parameters and reaction parameters for the different sediments are given in Table 9. The parameters listed are based on a broad literature survey of field-based determinations and experimental studies, combined to a comparative analysis of computed and measured depth profiles of reactive species. The latter approach is presently the sole way to obtain rate constants for many of the biogeochemical reactions taking place in natural sediments.

Computational results are presented in Figures 3 to 7 and in Table 10. The model correctly reproduces some of the major features of sedimentary biogeochemical dynamics. As observed in the real world, it is found that organic matter in deep-sea sediments is mostly degraded aerobically, while anaerobic degradation pathways, in particular sulfate reduction, dominate in organic-rich shelf and nearshore sediments [2]. Because of rapid depletion of oxygen combined with a limited availability of sulfate, most sedimentary organic matter is fermentatively degraded in sediments deposited in highly productive lakes [2].

By explicitly taking into account the reoxidation reactions of reduced species formed during the oxidation of organic matter, the model is capable of calculating what fraction of a given oxidant supplied to the sediment is directly utilized for the oxidation of organics and what fraction serves to oxidize secondary reduced species. This is illustrated in Table 10 which compares the relative importance of the various reactions reducing oxygen, Mn(III, IV) and Fe(III) in the sediments. It is apparent from the table that most dissolved oxygen consumed in deep-sea and oligotrophic lake sediments is utilized directly during

Table 10. Representative Sediments: (1) depth-integrated rates of organic carbon oxidation and relative distribution of the various organic carbon oxidation pathways (reactions A-1 to A-6), (2) total depth-integrated rates of oxygen, manganese(IV) and iron(III) reduction, plus % of each reduction process that is coupled directly to organic carbon oxidation, (3) recycling efficiencies of iron and manganese (equation 29 in text), and (4) oxygen penetration depths.

	Deep-Sea	Shelf	Coastal/ Estuarine	Oligotrophic lake	Eutrophic lake
C_{org} oxidation (μmol cm^{-2} a^{-1})	7	79	981	38	759
% distribution					
O_2	79.8	6.0	4.3	47.6	3.0
NO_3	10.7	5.8	1.6	9.9	9.3
$Mn(IV)$	2.9	0	0.1	0.2	0
$Fe(III)$	0.6	0.8	3.4	0.2	0.4
SO_4	6.0	87.3	90.6	4.7	0
C_{org}	0	0	0	37.3	87.3
O_2 reduction (μmol cm^{-2} a^{-1})	7	80	362	38	114
% C_{org} oxidation	93	7	14	59	21
Fe(III) reduction (μmol cm^{-2} a^{-1})	0.6	8	140	0.6	44
% C_{org} oxidation	29	34	95	50	27
Mn(IV) reduction (μmol cm^{-2} a^{-1})	0.7	4	66	0.2	19
% C_{org} oxidation	71	0	2	69	0
Recycling efficiencies (%)					
Fe	$\approx$100	99	93	97	63
Mn	$\approx$100	99	94	95	79
O_2 penetration (mm)*	105	12	8	25	2

* defined as the depth where the dissolved oxygen concentration drops below 1µM.

aerobic respiration. In the other cases, however, major fractions of porewater oxygen are diverted to the reoxidation of reduced species produced below the oxic surface layer. In fact, the calculations presented here suggest that this fraction dominates oxygen reduction in organic-rich shelf and coastal marine sediments, as well as in sediments of fertile freshwater environments.

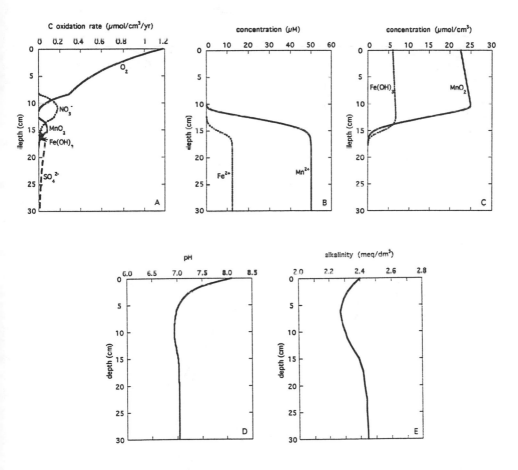

Figure 3.   *Deep-sea Sediment. Calculated depth profiles of (a) rates of organic carbon oxidation by aerobic respiration, denitrification, Mn reduction, Fe reduction, SO_4 reduction and methanogenesis, (b) concentrations of dissolved Mn^{2+} and Fe^{2+}, (c) concentrations of Mn and Fe oxides (expressed per unit total volume sediment), (d) porewater pH, and (e) total alkalinity (Equation 18 in text) and carbonate alkalinity (Equation 20 in text). Conditions: see Table 9.*

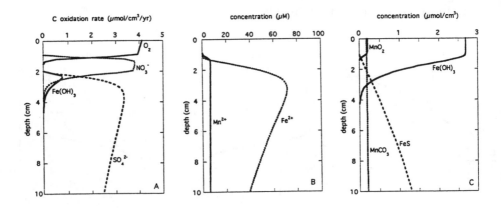

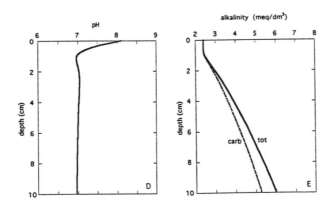

Figure 4.   *Continental Shelf Sediment.*

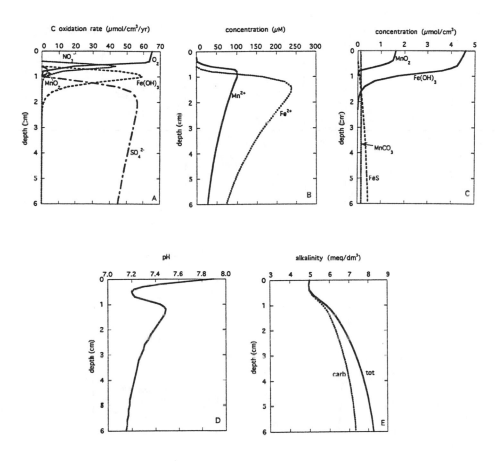

Figure 5. **Coastal-estuarine Sediment.**

2

METAL CYCLING IN SURFACE SEDIMENTS: MODELING THE INTERPLAY OF TRANSPORT AND REACTION

Philippe Van Cappellen and Yifeng Wang
School of Earth and Atmospheric Sciences
Georgia Institute of Technology
Atlanta, GA 30332

We are trained to think in terms of linear causality, but we need new "tools of thought": one of the greatest benefits of models is precisely to help us discover these tools and learn how to use them.

Ilya Prigogine and Isabelle Stengers [1]

1.   INTRODUCTION

Surface sediments are not the passive recipients of particulate metals settling out of the water column. Rather, they act as biogeochemical reactors in which the deposited metals participate in a variety of processes, including microbial reactions, redox transformations, adsorption-desorption exchanges, and the precipitation and dissolution of minerals. These processes regulate metal speciation and, therefore, control the return of metals to the overlying aquatic environment or their retention in the underlying sediment repository (Figure 1).

The accurate description of the distributions, transformations and transport of metal species in surface sediments requires a combination of approaches and methodologies. Field-based measurements offer an integrated record of the interaction of the sedimentological, biological and geochemical processes that affect metal cycling. Laboratory studies, on the other hand, focus on the mechanisms, kinetics and equilibrium states of individual biogeochemical processes. Mathematical sediment models form the bridge between field and experimental studies.

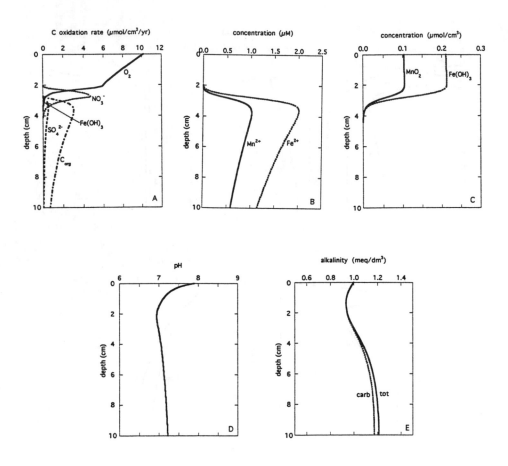

Figure 6.   *Oligotrophic Lake Sediment.*

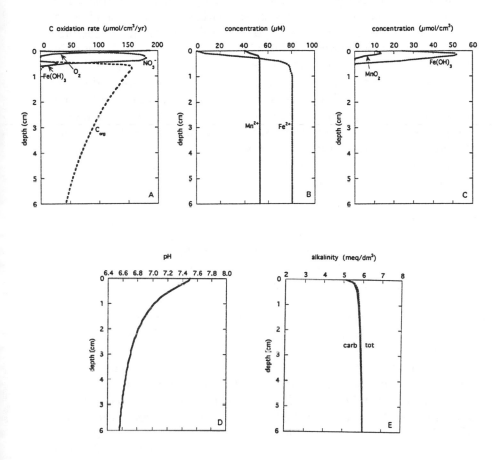

Figure 7.   Eutrophic Lake Sediment.

The results in Table 10 allow us to compare total oxygen consumption in a sediment to the integrated organic carbon oxidation rate. With the exceptions of the coastal-estuarine and the eutrophic lake sediments, the total oxygen consumption closely approximates the total oxidation of organic carbon taking place in the sediments. Thus, in many cases the uptake rate of O_2 at the water-sediment interface should provide a reliable measure of total carbon oxidation in the sediment. However, in depositional settings characterized by oxygen-poor bottom waters and/or very high rates of organic matter oxidation this agreement breaks down. Under these circumstances, significant portions of the reduced species released to the porewaters diffuse across the water-sediment interface and are reoxidized within the overlying water column.

Dissimilatory Fe(III) reduction (reaction A-4) competes with the reduction of ferric (hydr)oxides by upward diffusing porewater sulfide (reaction A-13). According to the calculations summarized in Table 10, the partitioning of Fe(III) (hydr)oxide reduction between the dissimilatory and non-dissimilatory pathways is highly variable from one sediment to another. This reflects the variability among the various depositional settings of the relative importance of Fe(III) and sulfate reduction, as well as the ratio of the rates of organic carbon and sulfide oxidation by Fe(III) (hydr)oxides. For example, almost all Fe(III) reduction in the coastal-estuarine sediment occurs via reaction A-4 (95%), because of the very high rate of organic matter oxidation which suppresses the non-dissimilatory reduction of Fe(III). It should be noted, however, that there are still large uncertainties associated with the values assigned to the rate constant of the reaction between sedimentary Fe(III) (hydr)oxides and porewater sulfide and, therefore, the calculated distributions of dissimilatory and non-dissimilatory Fe(III) reduction given here should be considered as tentative.

As for iron, the calculated partitioning of manganese reduction between dissimilatory and non-dissimilatory pathways varies greatly among the different model sediments. In the deep-sea and oligotrophic lake sediments most Mn reduction is coupled to organic carbon oxidation, because of relatively low rates of iron reduction and, consequently, little reaction between Fe^{2+} and manganese oxides phases (reaction A-9). In contrast, manganese oxides are mainly reduced by reaction with porewater Fe^{2+} in the three other sediments. Oxidation of Fe^{2+} by Mn oxides acts as a barrier to the upward diffusion of dissolved ferrous iron cations. This reaction explains the spatial separation between the rise of porewater Mn^{2+} and that of Fe^{2+} that is often observed in sediments [8,41,43].

The net addition of oxidative capacity to sediments through the deposition of metal oxides is usually small compared to the supply of dissolved external electron acceptors from the bottom waters or to the auto-oxidative capacity of the organic matter itself. However, because Fe and Mn undergo redox cycling in sediments covered by oxygen-containing bottom waters, the contribution of the metal oxides to the oxidation of sedimentary organic matter plus secondary reduced species may be much larger than would be predicted from the oxide deposition fluxes alone [41]. At steady state, a measure of the importance of redox cycling of Fe or Mn in a sediment is given by the recycling efficiency, E:

$$E_i = \frac{\int R_{ox,i} dx}{\int R_{red,i} dx} = 1 - \frac{F_i}{\int R_{red,i} dx} \qquad (29)$$

where the integrals represent the total, depth-integrated rates of oxidation or reduction of Mn and Fe, and F_i is the deposition flux of the metal oxide at the water-sediment interface. The value of E varies from 0 to 1. The lower boundary corresponds to the case where each deposited metal cation undergoes a single reduction before it is lost from the surface sediment, either through burial into the sediment repository or diffusion back to the water column. Non-zero values of E indicate that the metal cations are cycled several times among their redox states before being lost from the surface sediment. The average number of times an iron or manganese cation cycles through its oxidized and reduced forms is given by 1/(1-E).

Values for the recycling efficiencies of Fe and Mn in the different model sediments are given in Table 10. The lowest values of E are found in the eutrophic lake sediment, where the thin oxic surface layer cannot prevent the escape of significant fractions of the reduced metal cations to the overlying water column. In the other sediments, the recycling efficiencies are much higher, approaching 100% in shelf and deep-sea sediments. The somewhat lower values of E in the coastal-estuarine sediment are due to the relatively more important precipitation and burial of reduced iron and manganese mineral phases in this environment.

Redox cycling of iron and manganese within a sediment may cause a build-up of the oxide-bound metal concentrations in the oxidized portion of the sediment. This is most clearly illustrated by the deep-sea sediment where the oxic-anoxic boundary acts as a quasi-perfect trap for dissolved metal cations. In the absence of redox cycling, the concentrations of reactive Fe and Mn oxides in the top part of this sediment should not exceed a few hundredths of $\mu mol/cm^3$. The calculated concentrations, however, are several orders of magnitude higher (Figure 3b), because of the upward redistribution by particle mixing of metal oxides precipitated within the oxic to anoxic transition zone.

The concentrations and cycling of reactive Fe and Mn species in sediments depend on the magnitudes of particle and fluid transport. This is illustrated in Figure 8 where the recycling efficiencies and the maximum observed metal oxide concentrations in the coastal-estuarine sediment are plotted as a function of the particle mixing coefficient. The recycling efficiencies are only slightly affected by changes in the biodiffusion coefficient. Large differences, however, are observed in the absolute concentrations of metal oxides. Lower biodiffusion coefficients correspond to a less rapid removal of the oxides from the oxidation fronts and, hence, higher and sharper maxima in the metal oxide depth profiles.

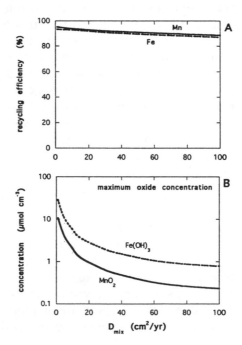

Figure 8.   *Effect of the Particle Mixing Coefficient on (a) the Recycling Efficiencies of Mn and Fe and (b) the Maximum Concentrations of Mn and Fe Oxides in the Coastal-estuarine Sediment. All other conditions are the same as in Figure 5.*

The simulations in Figure 8 emphasize two important features of sediment biogeochemical dynamics. First, in complex transport-reaction systems such as sediments there may not exist a simple relationship between the concentration of a given reactive species and its rate of biogeochemical transformation. Second, in order to predict the behavior, speciation and retention of metals in sediments, one requires a quantitative understanding of, not only the chemistry and biochemistry involved, but also the physical transport processes.

Porewater pH and alkalinity are sensitive indicators of sediment biogeochemistry. Figures 3 to 7 show that each model sediment has its own characteristic pH and alkalinity signature. Typically, the pH of the porewaters drops below the water-sediment interface, reflecting acid production as a result of O_2 reduction [44]. In the eutrophic lake sediment this decreasing trend persists below the aerobic layer as denitrification and methanogenesis take over as the major processes of organic matter degradation. In the coastal-estuarine and in the shelf sediment, however, the pH reaches a minimum at the base of the aerobic layer, followed by an increase. The observed pH maximum coincides with the alkalinity production during Mn and Fe reduction coupled to organic carbon and porewater sulfide oxidation (reactions A-3, A-4, A-12 and A-13). Below the maximum, porewater pH decreases again with depth. This decrease is driven mainly by the release of protons during the precipitation of hydrogen sulfide with ferrous iron (reaction A-19). For sufficiently low sulfide precipitation rates the pH may actually continue to rise with depth as observed, e.g., in [45].

The features of the pH profile, e.g., the pH minimum and maximum described above for the coastal-estuarine sediment, are sensitive to the relative importance of the various redox reactions taking place. For instance, comparatively more hydrogen ions are produced per O_2 molecule during the oxidation of Mn^{2+} and Fe^{2+} ions (reactions A-7 and A-8) than during aerobic respiration (reaction A-1). Thus, well-developed metal oxidation fronts in sediments may be expected to be associated with pronounced pH minima. Similarly, the magnitude and width of the pH maximum observed in Figure 7d is sensitive to the relative contributions of metal oxides and oxygen to the oxidation of sulfide diffusing up from the sulfate reduction zone.

As was pointed out several times during the previous discussion, sediment redox chemistry offers many examples of multiple pathways competing for the same species. The fact that these pathways often have different effects on porewater alkalinity and pH offers one potential avenue for evaluating the contributions of the different pathways (compare, for instance, the stoichiometries of the sulfide oxidation reactions A-11, A-12 and A-13). Technological advances in the last decade have produced microelectrodes that permit the measurement of depth profiles of pH and other porewater properties with spatial resolutions inferior to a millimeter. We foresee that the combination of micro-profiling of porewaters and modeling of the type presented in this paper will resolve many of the questions about the detailed biogeochemistry of surface sediments that until recently may have seemed intractable.

6.   CONCLUSIONS

The development of coupled, multi-component transport-reaction models that simulate the behavior of metals and associated reactive species in surface sediments has been made possible in the last decades by the rapidly increasing accessibility to powerful workstations and personal computers. In this paper, we have tried to show that despite the complexity of sediment biogeochemistry, the fundamental transport-reaction theory and its implementation into numerical computer codes is relatively straightforward. In fact, the construction of realistic biogeochemical models for surface sediments is presently limited by our ability to identify, formalize and parameterize the individual transport and reaction processes, rather than by inadequate computational and mathematical tools.

Acknowledgments: The authors acknowledge the fruitful discussions with Don Canfield on the cycling of Fe and Mn in sediments. They wish to thank Alakendra Roychoudhury for his help in developing the sediment model presented in this paper. Through their support, their willingness to share data and expertise, or by providing inspiration, the following people also deserve our gratitude: Bob Berner, Jean-François Gaillard and Tracey Tromp. This research was funded by the U.S. Environmental Protection Agency under Cooperative Agreement No. 820813, administered through the Environmental Research Laboratory, Athens, Georgia.

REFERENCES

1.   Prigogine, I. and I. Stengers, *Order out of Chaos: Man's New Dialogue with Nature* Bantam Books, (New York, 1984).

2. Van Cappellen, P., J.-F. Gaillard and C. Rabouille, "Biogeochemical Transformations in Sediments: Kinetic Models of Early Diagenesis," in *Interactions of C, N, P and S Biogeochemical Cycles and Global Change*, R. Wollast, F.T. Mackenzie and L. Chou, Eds., pp. 401-445 Springer-Verlag, (Berlin, 1993).

3. Middelburg, J.J., "A Simple Rate Model for Organic Matter in Marine Sediments," *Geochim. Cosmochim. Acta* 53:1577-1581 (1989).

4. Berner, R.A., *Early Diagenesis: A Theoretical Approach* Princeton Univ. Press, (Princeton, 1980).

5. Boudreau, B.P., "Mathematics of Tracer Mixing in Sediments: II. Nonlocal Mixing and Biological Conveyor-Belt Phenomena," *Amer. J. Sci.* 286:199-238 (1986).

6. Lapidus, L. and G.F. Pinder, *Numerical Solution of Partial Differential Equations in Science and Engineering* John Wiley and Sons, (New York, 1982).

7. Berner, R.A., "Sedimentary Pyrite Formation: An Update," *Geochim. Cosmochim. Acta* 48:605-615 (1984).

8. Wersin, P., P. Höhener, R. Giovanoli and W. Stumm, "Early Diagenetic Influences on Iron Transformations in a Freshwater Lake Sediment," *Chem. Geol.* 90:233-252 (1991).

9. Mucci, A., "Manganese Uptake during Calcite Precipitation from Seawater: Conditions Leading to the Formation of a Pseudokutnahorite," *Geochim. Cosmochim. Acta* 52:1859-1868 (1988).

10. Boudreau, B.P. and B.R. Ruddick, "On a Reactive Continuum Representation of Organic Matter Diagenesis," *Amer. J. Sci.* 291:507-538 (1991).

11. Burdige, D.J., "The Kinetics of Organic Matter Mineralization in Anoxic Marine Sediments," *J. Mar. Res.* 49:727-761 (1991).

12. Gaillard, J.-F. and C. Rabouille, "Using Monod Kinetics in Geochemical Models of Organic Carbon Mineralization in Deep-Sea Surficial Sediments," in *Deep-Sea Food Chains and the Global Carbon Cycle*, G.T. Rowe and V. Pariente, Eds., pp. 309-324 Kluwer Academic Publ., (The Netherlands, 1992).

13. Nielsen, L.P., P.B. Christensen, N.P. Revsbech and J. Sørensen, "Denitrification and Oxygen Respiration in Biofilms Studied with a Microsensor for Nitrous Oxide and Oxygen," *Microb. Ecol.* 19:63-72 (1990).

14. Billen, G., "A Budget of Nitrogen Recycling in North Sea Sediments Off the Belgian Coast," *Estuarine Coastal Mar. Sci.* 7:127-146 (1978).

15. Esteves, J.L., G. Mille, F. Blanc and J.C. Bertrand, "Nitrate Reduction Activity in a Continuous Flow-through System in Marine Sediments," *Microb. Ecol.* 12:283-290 (1986).

16. Murray, R.E., L.L. Parson and M.S. Smith, "Kinetics of Nitrate Utilization by Mixed Populations of Denitrifying Bacteria," *Appl. Environ. Microbiol.* 55:717-721 (1989).

17. Boudreau, B.P. and J.T. Westrich, "The Dependence of Bacterial Sulfate Reduction on Sulfate Concentration in Marine Sediments," *Geochim. Cosmochim. Acta* 48:2503-2516 (1984).

18. Ingvorsen, K., A.J.B. Zehnder and B.B. Jørgensen, "Kinetics of Sulfate and Acetate Uptake by *Desulfobacter postgatei*," *Appl. Environ. Microbiol.* 47:403-408 (1984).

19. Lovley, D.R., "Organic Matter Mineralization with the Reduction of Ferric Iron: A Review," *Geomicrobiol. J.* 5:375-399 (1987).

20. Morgan, J.J., W. Sung and A. Stone, "Chemistry of Metal Oxides in Natural Water: Catalysis of the Oxidation of Manganese(II) by γ-FeOOH and Reductive Dissolution of Manganese(III) and (IV) Oxides," in *Environmental Inorganic Chemistry*, K.J. Irgolic and A.E. Martell, Eds., pp. 167-184 VCH Publishers, Inc., (Weinheim, Germany 1985).

21. Wehrli, B., "Redox Reactions of Metal Ions at Mineral Surfaces," in *Aquatic Chemical Kinetics*, W. Stumm, Ed., pp. 311-336 John Wiley and Sons, (New York, 1990).

22. Emerson, S., S. Kalhorn, L. Jacobs, B.M. Tebo, K.H. Nealson and R.A. Rosson, "Environmental Oxidation Rate of Manganese(II): Bacterial Catalysis," *Geochim. Cosmochim. Acta* 46:1073-1079 (1982).

23. Kepkay, P.E., D.J. Burdige and K.H. Nealson, "Kinetics of Bacterial Manganese Binding and Oxidation in the Chemostat," *Geomicrobiol. J.* 3:245-262 (1984).

24. Tebo, B.M. and S. Emerson, "Effect of Oxygen Tension, Mn(II) Concentration, and Temperature on the Microbially Catalyzed Mn(II) Oxidation Rate in a Marine Fjord," *Appl. Environ. Microbiol.* 50:1268-1273 (1985).

25. Morse, J.W., F.J. Millero, J.C. Cornwell and D. Rickard, "The Chemistry of Hydrogen Sulfide and Iron Sulfide in Natural Waters," *Earth-Science Rev.* 24:1-42 (1987).

26. Pyzik, A.J. and S.E. Sommer, "Sedimentary Iron Monosulfides: Kinetics and Mechanism of Formation," *Geochim. Cosmochim. Acta* 45:687-698 (1981).

27. Yao, W. and F.J. Millero, "The Rate of Sulfide Oxidation by δMnO_2 in Seawater," *Geochim. Cosmochim. Acta* 57:3359-3365 (1993).

28. Moses, C.O., D.K. Nordstrom, J.S. Herman and A.L. Mills, "Aqueous Pyrite Oxidation by Dissolved Oxygen and by Ferric Iron," *Geochim. Cosmochim. Acta* 51:1561-1571 (1986).

29. Lerman, A., *Geochemical Processes: Water and Sediment Environments* John Wiley and Sons, (New York, 1979).

30. Ullman, W.J. and R.C. Aller, "Diffusion Coefficients in Nearshore Marine Sediments," *Limnol. Oceanogr.* 27:552-556 (1982).

31. Li, Y.-H and S. Gregory, "Diffusion of Ions in Sea Water and in Deep-Sea Sediments," *Geochim. Cosmochim. Acta* 38:703-714 (1974).

32. Tyson, R.V. and T.H. Pearson, "Modern and Ancient Continental Shelf Anoxia: An Overview," in *Modern and Ancient Continental Shelf Anoxia*, R.V. Tyson and T.H. Pearson, Eds., pp. 1-24 Geological Society Special Publication 58, (London, 1991).

33. Christensen, J.P., A.H. Devol and W.M. Smethie, Jr., "Biological Enhancement of Solute Exchange Between Sediments and Bottom Water on the Washington Continental Shelf," *Continental Shelf Res.* 3:9-23 (1984).

34. Boudreau, B.P., "On the Equivalence of Non-Local and Radial-Diffusion Models for Porewater Irrigation," *J. Mar. Res.* 42:731-735 (1984).

35. Emerson, S., R. Jahnke and D. Heggie, "Sediment-Water Exchange in Shallow Water Estuarine Sediments," *J. Mar. Res.* 42:709-730 (1984).

36. Stumm, W. and J.J. Morgan, *Aquatic Chemistry* John Wiley and Sons, (New York, 1981).

37. Mehrbach, C., C.H. Culberson, J.E., Hawley and R.M. Pytkowicz, "Measurement of the Apparent Dissociation Constants of Carbonic Acid in Seawater at Atmospheric Pressure," *Limnol. Oceanogr.* 18:897-907 (1973).

38. Millero, F.J., "The Thermodynamics of the Carbonate System in Seawater," *Geochim. Cosmochim. Acta* 43:1651-1661 (1979).

39. Goldhaber, M.B. and I.R. Kaplan, "Apparent Dissociation Constants of Hydrogen Sulfide in Chloride Solutions," *Marine Chem.* 3:83-104 (1975).

40. Millero, F.J., "The Thermodynamics and Kinetics of the Hydrogen Sulfide System in Natural Waters," *Marine Chem.* 18:121-147 (1986).

41. Canfield, D.E., B. Thamdrup and J.W. Hansen, "The Anaerobic Degradation of Organic Matter in Danish Coastal Sediments: Iron Reduction, Manganese Reduction, and Sulfate Reduction," *Geochim. Cosmochim. Acta* 57:3867-3883 (1993).

42. Tromp, T.K., P. Van Cappellen and R.M. Key, "A Global Model for the Early Diagenesis of Organic Carbon and Organic Phosphorus in Marine Sediments," *Geochim. Cosmochim. Acta* 59:1259-1284 (1995).

43. Froelich, P.N., G.P. Klinkhammer, M.L. Bender, N.A. Luedtke, G.R. Heath, D. Cullen, P. Dauphin, D. Hammond, B. Hartman and V. Maynard, "Early Oxidation of Organic Matter in Pelagic Sediments of the Eastern Equatorial Atlantic: Sub-oxic Diagenesis," *Geochim. Cosmochim. Acta* 43:1075-1090 (1979).

44. Boudreau, B.P., "Modelling the Sulfide-Oxygen Reaction and Associated pH Gradients in Porewaters," *Geochim. Cosmochim. Acta* 55:145-159 (1991).

45. Jahnke, R.A., "Early Diagenesis and Recycling of Biogenic Debris at the Seafloor, Santa Monica Basin, California," *J. Mar. Res.* 48:413-436 (1990).

3

TRACE METAL CHEMISTRY IN POREWATERS

George W. Luther, III
College of Marine Studies
University of Delaware
Lewes, DE 19958

1. INTRODUCTION

This chapter reviews trace metal chemistry of porewaters in freshwater and marine sediments. There are three main topics that will be described. These are the variety of sampling methods for obtaining porewaters, the processes which dissolve and precipitate metals, and the types of metal speciation that occur in porewaters. A recurrent theme in this discussion is "can we use porewater data to understand sedimentary processes"? The following discussion will present our current knowledge of the main topics and show where there are significant gaps. This paper will not describe the wealth of porewater metal data on total metal concentrations and their seasonal variations. The reader is referred to a few selected papers [1-3].

Because porewaters are characterized by sulfidic or low oxygen conditions, it is necessary to focus our discussion on anoxic and sulfidic processes.

2. EXPERIMENTAL

2.1 Sample handling

Prior to the discussion on sampling devices, it is necessary to point out that porewaters are characterized by low oxygen concentrations or zero oxygen. Therefore, all sampling methods and subsequent analyses must insure that oxygen is rigorously excluded from the sample and the extracted porewaters to prevent any oxidation artifacts. This requires the use of glove bags and/or glove boxes which are purged with the highest purity nitrogen or argon. Argon is recommended when the sample may need to sit for any length of time prior to analysis (e.g., during centrifugation) because it is heavier than air and less easily displaced than nitrogen by air. Vessels which are impermeable to air are also recommended. Although Teflon is recommended for trace metals since it can be easily cleaned of metals, Teflon is more permeable to air than most other plastics. In the author's laboratory, polypropylene has been used effectively without trace metal and air contamination of the sample.

2.2   Common porewater sampling devices

There are six common devices used in porewater sampling and analysis. These are the Reeburgh squeezer, centrifugation, peepers, wells, whole-core squeezers and microelectrodes. All have advantages and disadvantages depending on the sediment system under investigation [4].

The Reeburgh squeezer [5] and centrifugation [6] rely on the procurement of cores which must be sectioned into 0.5 cm or greater intervals. The sections are cut or scraped with clean plasticware, then processed for porewater by loading the sediment into the appropriate device. For the Reeburgh squeezer, the sediment is placed into a chamber with a diaphragm which can be pressurized with nitrogen. The nitrogen presses against the diaphragm which in turn presses against the sediment with the porewater being expressed out the other end of the chamber. The exit end can be set up with appropriate filters, usually 0.4 μm, to remove the fine sediments from the porewaters. The porewater should be collected in a syringe to prevent any outgassing of gases such as H_2S.

In centrifugation, the sediment is placed into a centrifuge tube and centrifuged at 5,000 to 11,000 rpm for fine sediments [7] and less for coarse grain sediments [8]. The water is quickly removed and filtered through 0.40 μm filters before the porewater can be reabsorbed by the sediments. In the centrifugation method, significant amounts of gas will be released into the tube unless the tube is filled with sediment to the top. The centrifugation method does not usually provide as much porewater as the Reeburgh squeezer. However, the Reeburgh squeezer can provide more interstitial water from plants and their roots which is not part of the porewater. The material being expressed from the plants is usually of an organic nature [9]. Both of these methods do not allow for vertical resolution better than 0.25 cm. This is a problem in many sediments where biogeochemical changes occur at submillimeter vertical resolution [10,11].

Peepers are usually dialysis membranes or gels which obtain porewater by equilibration of the fluid in the membrane or gel with the porewater [12]. The peeper is deployed into the sediment and is allowed to equilibrate with the sediment for periods of 10 days or more; peepers containing gels have shorter equilibration times. The porewater is then sampled for analysis or the gel is sectioned for analysis. Recently, in the case of gels, the gel has been analyzed by solid state methods (e.g., PIXE) at millimeter resolution and has provided the first information on metal profiles such as Fe at this fine scale vertical resolution [13]. Because peepers require significant equilibration times, they are not as useful as other methods for very short term temporal studies [9,14]. However, centrifugation and dialysis membranes do give similar data for trace metals [7] such as Co, Ni, Cr, Fe, Mn when short term environmental changes do not occur.

Whole-core squeezers are popular in the marine field [15,16]. Here, sediments are obtained by traditional box coring methods and the box is subsampled onboard ship with a plastic core tube. This core tube can be machined with inlets for the introduction of syringes at about 0.5 cm intervals. The syringes are attached to the core barrel with a Luer lock fitting which had been screwed into the barrel. Porex tubing with a nominal 100 μm filtering capacity is attached to the inside of the Luer lock portion of the syringe. The top and the bottom of the core are enclosed with the top having an inlet for the introduction of nitrogen gas. The gas is added to pressurize the top of the sediment. As the sediment is pressurized, the porewaters flow into the syringes which are later filtered to remove fine grain particles. A disadvantage of this method can be the compaction of the sediment during nitrogen pressurization.

Wells [17] and other *in situ* variants [18,19] are deployed much like peepers. However, suction is used to obtain the sample. An accurate determination of the vertical dimension at millimeter resolution is not always possible. However, much porewater can be obtained and it is not contaminated with organic material as in the cutting methods (Reeburgh squeezer and centrifugation).

Lastly, microelectrodes are now available for the determination of a variety of species [10,11,20-23]. Microelectrodes can be membrane or solid state. Membrane types allow gases or H^+ to diffuse through to an inner electrode system which can be either potentiometric or polarographic in design. Potentiometric microelectrodes are now available for H^+ and pCO_2 [20] whereas polarographic microelectrodes are available for O_2, H_2S [21,22] and some nitrogen gases [23]. Microelectrodes have the advantage of being non-destructive to the environment when they are deployed in the field. Although, not all the desired species can be determined by them, recent developments indicate that the number of species which can be determined by them simultaneously is increasing [22]. Membrane type electrodes are susceptible to breakage and do not allow for the transport of trace metal ions or complexes across the membrane for measurement. However, solid state electrodes are more robust to breakage and do allow for the direct measurement of metal ions and other non-gaseous analytes. These solid state electrodes require electrochemical conditioning between each measurement and can also determine O_2 and H_2S. In our laboratory, we have developed solid state microelectrodes that can measure O_2, H_2S, Fe and Mn using voltammetric techniques. Figure 1 shows data on Fe in the top few mm of a salt marsh core [24].

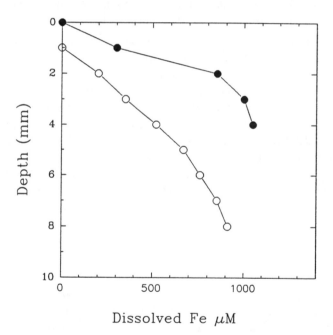

Dissolved Fe μM

Figure 1.   *Profile of Iron in a Salt Marsh Core from Great Marsh, Delaware during June 1992 Using a Solid State Electrode [24]. Oxygen and sulfide were not detected below the surface of the core. Filled symbols are analyses performed on the day of sampling; open symbols were performed a day later and show the effect of oxygen penetration into the core.*

3. MINERAL FORMATION AND DISSOLUTION PROCESSES

3.1 Mineral formation

The most common minerals which contain trace metals are the clays. However, these are typically resistant to attack by microbes and chemicals including strong acids. Thus, they do not appear to be as important in trace metal cycling as the oxides and oxyhydroxides of iron and manganese and the sulfides of iron. The oxides and oxyhydroxides of iron and manganese readily form during oxidation by oxygen in the environment [25]. Iron(II) readily oxidizes at lower pH than manganese(II). Inorganic Fe(II) oxidizes at pH values of > 5 whereas inorganic Mn(II) does not oxidize readily until pH values > 8.5. Both of these reactions are autocatalyzed by the formation of their own oxide phases.

Sulfide phases are formed in anoxic and sulfidic environments when sulfide attacks and dissolves the oxide and oxyhydroxide phases. This process reduces Fe(III) to Fe(II) which precipitates FeS and FeS_2. The formation of pyrite, FeS_2, requires zerovalent sulfur, S(0), which is found as elemental sulfur and polysulfides [26] in marine environments and which are formed during the reduction of the oxide minerals [27-29].

The oxide, oxyhydroxide and sulfide phases are efficient traps for other trace metals through adsorption and coprecipitation [30]. Morse [31] shows data on the incorporation of trace metals in these phases found in a host of marine systems. Upon dissolution of any of these phases, other trace metals can be released to porewaters. Thus, an understanding of the chemistry and cycling of these solid phases is essential to the understanding of trace metals in porewaters.

3.2 Mineral dissolution

The following calculation indicates the importance of mineral dissolution to the release of dissolved metal to porewaters. For a typical salt marsh sediment of density 1.8 g/mL and total mineral concentration of 1.0 μmole/g dry wt., the total metal released to the porewaters on dissolution would be 1.8 μmole/mL or 1.8 mM. Typically the iron and manganese solid phases occur in concentrations ranging from 10 to 200 μmole/g dry wt. Thus, only a relatively small change in the solid phase through dissolution can have profound effects on porewater concentrations. Mineral dissolution can occur by several processes which include bacterial decomposition of organic matter, chemical reduction, chemical oxidation and non-redox initiated dissolution. Each of these will be discussed below.

During bacterial decomposition of organic matter in sediments, the oxidants consumed begin with the thermodynamically favorable and proceed through the least thermodynamically favorable until all the oxidants are consumed [32]. As shown in (Table 1), manganese oxide reduction should occur readily and in fact does because oxygen and nitrate are not found in high concentrations relative to the oxide phases. Although iron oxide reduction appears less favorable, there is overwhelming evidence showing that it is an important process. This reflects the higher relative concentration of iron oxide phases over manganese oxide phases in most environments. The succession of oxidants for bacterial decomposition of organic matter indicates that there are discrete zones in the sediments' porewaters which show buildup and depletion of Mn(II) followed by buildup and depletion of Fe(II) deeper in the sediment. Although this occurs for deep ocean sediments [30], it is not usually seen in freshwater lakes [33] and salt marsh environments [26] where sulfate reduction appears to be the dominant process even in surface sediments.

Abiotic reductive mineral dissolution can occur for both manganese and iron oxide phases (Table 2). The primary reductants are H_2S [27-29] and organic compounds such as phenols and catechols [34,35] which are breakdown products of organic matter. Iron oxide phases undergo another reductive dissolution process which manganese does not (Equation 3, Table 2). This process is the autocatalyzed dissolution of Fe(III) minerals by Fe(II) organic complexes whereby the total iron and Fe(III) concentration increases but the Fe(II) concentration remains constant [36,37]. A detailed mechanism [37] for this reaction has been proposed.

Non-redox initiated dissolution of the iron oxide phases occurs by acid which can be generated by other processes (e.g., oxidation of sulfide and pyrite) and by organic ligands which are breakdown products from organic matter decomposition (Equation 5, Table 2). Many organic ligands can solubilize iron oxide phases with formation of strong Fe(III) complexes. The siderophores are a prime example of this process [38].

For the sulfides of iron, two processes are important in their dissolution. The first is the proton assisted dissolution of FeS. The second is the oxidative dissolution of FeS_2 which generates substantial acidity. Pyrite does not dissolve in acid because it is kinetically inert to dissociation as a result of the low spin state (t_{2g}^6) for Fe(II).

Table 1.	Oxidation Reactions for the Decomposition of Organic Matter. Gibbs Free Energies are given per mole of glucose (adapted from [32]).	
		ΔG (kJ/mole)
OXYGEN REDUCTION $138 \, O_2 + C_{106}H_{263}O_{110}N_{16}P \rightarrow$ $\quad 106 \, CO_2 + 16 \, HNO_3 + H_3PO_4 + 122 \, H_2O$		-3190
NITRATE REDUCTION $94.4 \, HNO_3 + C_{106}H_{263}O_{110}N_{16}P \rightarrow$ $\quad 106 \, CO_2 + 55.2 \, N_2 + H_3PO_4 + 177.2 \, H_2O$		-3030
MnO_2 REDUCTION $236 \, MnO_2 + C_{106}H_{263}O_{110}N_{16}P + 472 \, H^+ \rightarrow$ $\quad 8 \, N_2 + 106 \, CO_2 + H_3PO_4 + 236 \, Mn^{2+} + 366 \, H_2O$		-3000
Fe_2O_3 REDUCTION $212 \, Fe_2O_3 + C_{106}H_{263}O_{110}N_{16}P + 848 \, H^+ \rightarrow$ $\quad 106 \, CO_2 + 16 \, NH_3 + H_3PO_4 + 424 \, Fe^{2+} + 530 \, H_2O$		-1410
SULFATE REDUCTION $53 \, SO_4^{2-} + C_{106}H_{263}O_{110}N_{16}P \rightarrow$ $\quad 106 \, CO_2 + 16 \, NH_3 + H_3PO_4 + 53 \, S^{2-} + 106 \, H_2O$		- 380
METHANOGENESIS $C_{106}H_{263}O_{110}N_{16}P \rightarrow$ $\quad 53 \, CO_2 + 53 \, CH_4 + 16 \, NH_3 + H_3PO_4$		- 350

$$FeS + H^+ \rightarrow Fe^{2+} + HS^-$$

$$FeS_2 + 14 \, Fe^{3+} + 8 \, H_2O \rightarrow 15 \, Fe^{2+} + 2 \, SO_4^{2-} + 16 \, H^+$$

In contrast, Fe(II) in FeS is labile to dissociation due to the high spin state $(t_{2g}^4 \, e_g^2)$ for Fe(II). Metals in the first transition series show both high and low spin states depending on the ligand, with the low spin state having stronger bonds and extra stability due to ligand field stabilization. Thus, for pyrite to dissolve, the disulfide ligand in pyrite must be attacked by oxidation. The best environmental oxidant for pyrite is dissolved Fe(III), not oxygen [26,39-42]. The following section discusses dissolved metal speciation in porewaters which is important to mineral dissolution and formation processes.

Table 2.   FeOOH Dissolution Reactions Which are Base Generating or Acid Consuming Reactions.

<u>Reductive dissolution</u>

(1) $2\ FeOOH + H_2S \rightarrow 2\ Fe^{2+} + S^0 + 4\ OH^-$

(2)* $2\ FeOOH + C_6H_4(OH)_2 \rightarrow 2\ Fe^{2+} + 4\ OH^- + C_6H_4(O)_2$
                     catechol   quinone

(3)+ $FeOOH + H^+ + Fe^{2+}(LL) \rightarrow Fe^{3+}(LL) + Fe^{2+}(LL) + 2\ OH^-$

<u>Non-reductive dissolution</u>

(4) $FeOOH + 3\ H^+ \rightarrow Fe^{3+} + 2\ H_2O$

(5) $FeOOH + C_6H_4(OH)_2 \rightarrow \{Fe^{3+}[C_6H_4(O)_2]\}^+ + OH^- + H_2O$
                     catechol

---

*Compounds with catechol functional groups dissolve iron minerals reductively at lower pH (< 5).

+LL = bidentate ligand (e.g., oxalate).

4.   FACTORS GOVERNING METAL SPECIATION IN POREWATERS

The predominant factors governing dissolved metal speciation in porewaters and metal reactivity are metal oxidation state and metal complexes including inorganic and organic ligands. As shown above for FeS and FeS_2, metal spin state also has a pronounced effect on metal reactivity. Because most porewater studies are performed with only 0.40 μm filtration, colloidal materials can pass through the filter. Thus, both FeS and hydrated Fe(III) oxides found in fresh and saline waters are not efficiently trapped by these filters. The following section reviews what is known about metal speciation in porewaters. In fact, little is known about metal speciation from experimental research primarily because of the small sample volumes obtained for analysis. For metals other than Fe and Mn, there are also low dissolved concentrations which also hamper detailed speciation information. Lastly, because of dramatic changes which can occur on the millimeter and submillimeter vertical scale, it is difficult to know whether the speciation has been changed by the sampling methodology - an environmental application for the Heisenberg uncertainty principle.

4.1   Metal oxidation states

At present, there are colorimetric reagents for the determination of Fe(II). The most commonly used are ferrozine and o-phenanthroline. Total iron can be determined by atomic absorption (AA) spectroscopy and inductively coupled plasma emission spectroscopy (ICP) or by reduction of Fe(III) phases followed by the determination of the resulting Fe(II) by colorimetric reagents [43]. Fe(III) is determined by difference between the total Fe and the Fe(II) measurements. Although there are colorimetric measurements for the determination of Mn(III), they have not been applied to environmental water analyses on a routine basis [44]. Polarography gives reduction peaks or waves for Mn(II) to Mn(0) and Fe(II) to Fe(0) at -1.5 V and -1.3 V, respectively, and these have been considered to be specific for these oxidation states [45]. However, Luther et al. [46] have shown that organic complexes for all Mn oxidation states (II,III,IV) give the Mn(II) reduction peak at the same potential, whereas the Mn(III) to Mn(II) reduction occurs over a wide range of potential. Thus, polarographic analysis using the Mn(II) reduction provides information on total metal only. Further work on the Fe system [47] indicates that the Fe(II) peak gives total metal concentration also. Thus, only colorimetric methods have been documented to give information on metal oxidation state.

Luther et al. [26] have shown that both Fe(II) and Fe(III) can coexist in salt marsh porewaters. Although the pH and organic content of the porewaters would allow for dissolved Fe(III), they could not indicate whether the Fe(III) was colloidal or dissolved because of the 0.40 μm operational filtration procedure used in sampling. Recently, Liang et al. [17] studied a sandy aquifer and filtered their samples through 1.5 nm filters with a molecular weight cutoff of < 3000. Up to 80% of the Fe(III) was determined to be complexed in these low oxygen waters. However, no direct evidence was provided for organic complexation of Fe(III) by this method.

4.2   Metals in organic complexes

There are only a few examples of experimental research which have documented organic complexation of metals in porewaters. In anoxic marine porewaters, Krom and Sholkovitz [48] used ultrafiltration membranes to fractionate metals into different classes. They found that most of the iron was complexed; however, the manganese was only complexed to a maximum of 12%. Lyons et al. [49] used XAD-2 resin to separate organic from inorganic iron and found iron to be complexed with organics to a maximum of 31%. In salt marsh porewaters, Giblin *et al.* [50] used Sephadex gels with molecular weight cutoffs up to 10,000 to separate organic from inorganic complexes. The method separates high molecular weight from low molecular weight complexes well; however, the low molecular weight complexes consist of both inorganic and organic forms. In their study, 34 -50% of the iron, 25 - 92% of the copper and only up to 9% of the manganese and cadmium were complexed to the high molecular weight material. It is likely that more of these metals were complexed to organics in the low molecular weight fraction because this fraction contains both inorganic and organic compounds. Based on these results with XAD-2 resin and Sephadex gels, it appears that ultrafiltration membranes [17,48] may overestimate the amount of organically complexed metals.

4.3   Metals in inorganic complexes

The determination of inorganic metal complexes is usually based on thermodynamic modeling [50-52]. These studies predict metal complexes with chloride, hydroxide, carbonate, sulfide and polysulfide. It is difficult to determine by experimental methods the chloro, hydroxo and carbonato complexes of metals. However, there is indirect evidence for sulfide complexes with Fe(II) because Fe(II) and sulfide coexist in many porewaters [50-52]. In several instances, Fe(II) and sulfide are supersaturated with respect to FeS precipitation. However, sulfide complexes can be determined by polarographic methods which are very sensitive to sulfide species [45,53]. Luther and Ferdelman [53] have shown that the sulfide measured in porewaters undergoes changes in the peak potential of the sulfide relative to sulfide standards prepared without metal in solution. Luther and Ferdelman [53] prepared Fe(II) and sulfide solutions to compare with their porewater sulfide data. They found that Fe(II) and sulfide form two complexes, $Fe(SH)^+$ and $Fe_2(SH)^{3+}$ with log stability constants of 5.5 and 11.1, respectively. Both complexes are kinetically labile because the sulfide is removed from solution on acidification and purging. The $Fe_2(SH)^{3+}$ complex is a multinuclear metal complex which should be important in pyrite synthesis. The two synthetic complexes are not as stable as the sulfide complex that they observed in creeks at low tide. These creek waters have significant amounts of porewater which drains into the creeks from salt marsh sediments at the creek bank. That complex was kinetically inert to dissociation by acid because, at a pH of 1.5 and 10 minutes of purging with argon, the sulfide was not removed from solution. The natural complex may be kinetically stabilized by organic complexation or by a low spin state for Fe(II) as in the case of pyrite.

5.   CONCLUSION

Porewater chemistry can indicate information about sedimentary processes; however, information on the solid state is also essential because small changes in the sediment cause large porewater changes. More information is needed on all aspects of metal speciation in porewaters, in particular, redox states and inorganic and organic complexation of metals. Metal speciation affects the metals' reactivity with both soluble and solid phase components. However, analytical methods for speciation are further complicated by the fact that metal concentrations and speciation change at the millimeter and submillimeter vertical scale in sediments. Thus, new sampling methods and strategies are needed in addition to better analytical methods. The choice of sampling and analytical methods for porewater work is dictated by a combination of sample regime and the types of questions which need to be addressed.

Acknowledgments: This work was supported by grants from the Ocean Sciences Division of NSF and the Office of Sea Grant (NOAA). I thank P. Brendel for the solid state microelectrode data shown in Figure 1.

REFERENCES

1.   Hines, M.E., W.B. Lyons, P.B. Armstrong, W.H. Orem, M.J. Spencer and H.E. Gaudette, "Seasonal Metal Remobilization in the Sediments of Great Bay, New Hampshire," *Mar. Chem.* 15:173-187 (1984).

2.   Giblin, A.E. and R.W. Howarth, "Porewater Evidence for a Dynamic Sedimentary Iron Cycle in Salt Marshes," *Limnol. Oceanogr.* 29:47-63 (1984).

3.   Gaillard, J.F., C. Jeandel, G. Michard, E. Nicolas and R. Renard, "Interstitial Water Chemistry of Villefranche Bay Sediments: Trace Metal Diagenesis," *Mar. Chem.* 18:233-247 (1986).

4.   Bufflap, S.E. and H.E. Allen, "Sediment Pore Water Collection Methods for Trace Metal Analysis: A Review," *Wat. Res.* 29:165-177 (1995).

5.   Reeburgh, W.S., "An Improved Interstitial Water Sampler," *Limnol. Oceanogr.* 12:163-165 (1967).

6. Emerson, S., V. Grundmanis and D. Graham, "Early Diagenesis in Sediments from the Eastern Equatorial Pacific. 1. Pore Water Nutrient and Carbonate Results," *Earth Planet. Sci. Lett.* 43:57-80 (1980).

7. Carignan, R., F. Rapin and A. Tessier, "Sediment Porewater Sampling for Metal Analysis: A Comparison of Techniques," *Geochim. Cosmochim. Acta* 49:2493-2497 (1985).

8. Saager, P.M., J. Sweerts and H.J Ellermeijer, "A Simple Pore-Water Sampler for Coarse, Sandy Sediments of Low Porosity," *Limnol. Oceanogr.* 35:747-751 (1990).

9. Howes, B.L., J.W.H. Dacey and S.G. Wakeham, "Effects of Sampling Technique on Measurements of Porewater Constituents in Salt Marsh Sediments," *Limnol. Oceanogr.* 30:221-227 (1985).

10. Revsbech, N.P., J. Sorensen, T.H. Blackburn and J.P. Lomholt, "Distribution of Oxygen in Marine Sediments Measured with Microelectrodes," *Limnol. Oceanogr.* 25:403-411 (1980).

11. Revsbech, N.P. and B.B. Jorgensen, "Microelectrodes: Their Use in Microbial Ecology," *Adv. Microb. Ecol.* 9:293-352 (1986).

12. Mayer, L.M., "Chemical Water Sampling in Lakes and Sediments with Dialysis Bags," *Limnol. Oceanogr.* 21:909-912 (1976).

13. Davison, W., G.W. Grime, J.A.W. Morgan and K. Clarke, "Distribution of Dissolved Iron in Sediment Pore Waters at Submillimetre Resolution," *Nature* 352:323-324 (1991).

14. Hesslein, R.H., "An In Situ Sampler For Close Interval Pore Water Studies," *Limnol. Oceanogr.* 21:912-914 (1976).

15. Bender, M., W. Martin, J. Hess, F. Sayles, L. Ball and C. Lambert, "A Whole Squeezer for Interfacial Pore-Water Sampling," *Limnol. Oceanogr.* 32:1214-1225 (1987).

16. Jahnke, R.A., "A Simple, Reliable, and Inexpensive Pore-Water Sampler," *Limnol. Oceanogr.* 33:484-487 (1988).

17. Liang, L., J.F. McCarthy, L.W. Joolet, J.A. McNabb and T.L. Mehlhorn, "Iron Dynamics: Transformation of Fe(II)/Fe(III) During Injection of Natural Organic Matter in a Sandy Aquifer," *Geochim. Cosmochim. Acta* 57:1987-1999 (1993).

18. Sayles, F.L., T.R.S. Wilson, D.N. Hume and P.C. Mangelsdorf, "*In Situ* Sampler for Marine Sedimentary Pore Waters: Evidence for Potassium Depletion and Calcium Enrichment," *Science* 181:154-156 (1973).

19. Watson, P.G. and T.E. Frickers, "A Multilevel, *In Situ* Pore-Water Sampler for Use in Intertidal Sediments and Laboratory Microcosms," *Limnol. Oceanogr.* 35:1381-1389 (1990).

20. Cai, W.J. and C.E. Reimers, "The Development of pH and pCO_2 Microelectrodes for Studying the Carbonate Chemistry of Pore Waters Near the Sediment - Water Interface," *Limnol. Oceanogr.* 38:1762-1773 (1993).

21. Revsbech, N.P., B.B. Jorgensen, T.H. Blackburn and Y. Cohen, "Microelectrode Studies of the Photosynthesis and O_2, H_2S, and pH Profiles of a Microbial Mat," *Limnol. Oceanogr.* 28:1062-1074 (1983).

22. Visscher, P.T., J. Beukema and H. van Gemerden, "*In Situ* Characterization of Sediments: Measurements of Oxygen and Sulfide Profiles with a Novel Combined Needle Electrode," *Limnol. Oceanogr.* 36:1476-1480 (1991).

23. Revsbech, N.P., L.P. Nielsen, P.B. Christensen and J. Sorensen, "Combined Oxygen and Nitrous Oxide Microsensor for Denitrification Studies," *Appl. Environ. Microbio.* 54:2245-2249 (1988).

24. Brendel, P. and G.W. Luther, III, "Development of a Gold Amalgam Voltammetric Microelectrode for the Determination of Dissolved Fe, Mn, O_2, and S(-II) in Porewaters of Marine and Freshwater Sediments," *Environ. Sci. Technol.* 29:751-761 (1995).

25. Stumm, W. and J.J. Morgan, *Aquatic Chemistry* John Wiley and Sons, (New York, 1981).

26. Luther, G.W., III, T.G. Ferdelman, J.E. Kostka, E.J. Tsamakis and T.M. Church, "Temporal and Spatial Variability of Reduced Sulfur Species (FeS_2, $S_2O_3^{2-}$) and Porewater Parameters in Salt Marsh Sediments," *Biogeochem.* 14:57-88 (1991).

27. Burdige, D.J. and K.H. Nealson, "Chemical and Microbiological Studies of Sulfide-Mediated Manganese Reduction," *Geomicrobio. J.* 4:361-387 (1986).

28. Pyzik, A.J. and S.E. Sommer, "Sedimentary Iron Monosulfides: Kinetics and Mechanism of Formation," *Geochim. Cosmochim. Acta* 45:687-698 (1981).

29. dos Santos Afonso, M. and W. Stumm, "Reductive Dissolution of Iron(III) (Hydr)oxides by Hydrogen Sulfide," *Langmuir* 8:1671-1675 (1992).

30. Stumm, W., *Chemistry of the Solid-Water Interface* John Wiley and Sons, (New York, 1992),

31. Morse, J.W., "Dynamics of Trace Metal Interactions with Authigenic Sulfide Minerals in Anoxic Sediments," in *Metal Speciation and Contamination of Aquatic Sediments*, H.E. Allen, Ed., pp. 187-199. Ann Arbor Press, (Chelsea, MI 1995).

32. Froelich, P.N., G.P. Klinkhammer, M.L. Bender, N.A. Luedtke, G.R. Heath, D. Cullen, P. Dauphin, D. Hammond, B. Hartman and V. Maynard, "Early Oxidation of Organic Matter in Pelagic Sediments of the Eastern Equatorial Atlantic: Suboxic Diagenesis," *Geochim. Cosmochim. Acta* 43:1075-1090 (1979).

33. Wersin, P., P. Hohener, R. Giovanoli and W. Stumm, "Early Diagenetic Influences on Iron Transformations in a Freshwater Lake Sediment," *Chem. Geol.* 90:233-252 (1991).

34. Lakind, J.S. and A.T. Stone, "Reductive Dissolution of Goethite by Phenolic Reductants," *Geochim. Cosmochim. Acta* 53:961-971 (1989).

35. Stone, A.T. and H. Ulrich, "Kinetics and Reaction Stoichiometry in the Reductive Dissolution of Manganese(IV) Oxide and Co(III) Oxide by Hydroquinone," *J. Coll. Interface Sci.* 132:509-522 (1989).

36. Sulzberger, B, D. Suter, C. Siffert, S. Banwart and W. Stumm, "Dissolution of Fe(III) (Hydr)oxides in Natural Waters; Laboratory Assessment on the Kinetics Controlled by Surface Coordination," *Marine Chem.* 28:127-144 (1989).

37. Luther, III, G.W., J.E. Kostka, T.M. Church, B. Sulzberger and W. Stumm, "Seasonal Iron Cycling in the Marine Environment: the Importance of Ligand Complexes with Fe(II) and Fe(III) in the Dissolution of Fe(III) Minerals and Pyrite, Respectively," *Marine Chem.* 40:81-103 (1992).

38. Hider, R.C., "Siderophore Mediated Absorption of Iron," *Structure and Bonding* 58:26-87 (1984).

39. Luther, III, G.W., "Pyrite Oxidation and Reduction: Molecular Orbital Therory Considerations," *Geochim. Cosmochim. Acta* 51:3193-3199 (1987).

40. Singer, P.C. and W. Stumm, "Acid Mine Drainage - The Rate Limiting Step," *Science* 167:1121-1123 (1970).

41. Moses, C.O., D.K. Nordstrom, J.S. Herman and A.L. Mills, "Aqueous Pyrite Oxidation by Dissolved Oxygen and by Ferric Iron," *Geochim. Cosmochim. Acta* 51:1561-1571 (1987).

42. Moses, C.O. and J.S. Herman, "Pyrite Oxidation at Circumneutral pH," *Geochim. Cosmochim. Acta* 55:471-482 (1991).

43. Stookey, L.L., "Ferrozine - A New Spectrophotometric Reagent for Iron," *Anal. Chem.* 41:779-781 (1970).

44. Ishii, H., H. Koh and K. Satoh, "Spectrophotometric Determination of Manganese Utilizing Metal Ion Substitution in the Cadmium $\alpha,\beta,\gamma,\delta$-tetrakis(4-carboxyphenyl)porphine Complex," *Anal. Chim. Acta* 136:347-352 (1982).

45. Davison, W., J. Buffle and R. De Vitre, "Direct Polarographic Determination of O_2, Fe(II), Mn(II), S(II) and Related Species in Anoxic Waters," *Pure Appl. Chem.* 60:1535-1548 (1988).

46. Luther, III, G.W., D. Nuzzio and J. Wu, "Speciation of Manganese in Chesapeake Bay Waters by Voltammetric Methods," *Anal. Chim. Acta*, 284:473-480 (1994).

47. Taylor, S.W., G.W. Luther, III and J.H. Waite, "Polarographic and Spectrophotometric Investigation of Iron(III) Complexation to 3,4-Dihydroxyphenylalanine-containing Peptides and Proteins from *Mytilus edulis*," *Inorg. Chem.* 33:5819-5824 (1994).

48. Krom, M.D. and E.R. Sholkovitz, "Nature and Reactions of Dissolved Organic Matter in the Interstitial Waters of Marine Sediments," *Geochim. Cosmochim. Acta* 41:1565-1573 (1976).

49. Lyons, W.B., H.E. Gaudette and P.B. Armstrong, "Evidence for Organically Associated Iron in Nearshore Pore Fluids," *Nature* 282:202-203 (1979).

50. Giblin, A.E., I. Valiela and G.W. Luther, III, "Trace Metal Solubility in Salt Marsh Sediments Contaminated with Sewage Sludge," *Estuar. Coastal Shelf Sci.* 23:477-498 (1986).

51. Gardner, L.R., "Organic Versus Inorganic Trace Metal Complexes in Sulfidic Marine Waters - Some Speculative Calculations Based on Available Stability Constants," *Geochim. Cosmochim. Acta* 38:1297-1302 (1974).

52. Boulegue, J, C.J. Lord and T.M. Church, "Sulfur Speciation and Associated Trace Metals (Fe,Cu) in the Porewaters of Great Marsh, DE," *Geochim. Cosmochim. Acta* 46:453-464 (1982).

53. Luther, III, G.W. and T.G. Ferdelman, "Voltammetric Characterization of Iron (II) Sulfide Complexes in Laboratory Solutions and in Marine Waters and Porewaters," *Environ. Sci. Technol.* 27:1154-1163 (1993).

METAL ADSORPTION ONTO AND DESORPTION FROM SEDIMENTS: I. RATES

Everett A. Jenne
Geosciences Department
Battelle, Pacific Northwest Laboratories
Richland, WA 99352

1.   INTRODUCTION

Establishing criteria for a maximum allowable metal content of sediments that will not cause chronic toxicity to aquatic organisms requires a science-based modeling capacity [1-3]. Understanding the availability of metals to aquatic organisms and the cycling of nutrient metals in the marine environment requires understanding the rates of adsorption and desorption processes [4-8]. Sorption rates, $R_{Me,ads}$ or $R_{Me,des}$, provide a means of estimating and modeling the contribution of adsorbed sediment to aquatic organism uptake of metals from sediment as distinct from metals dissolved in the water [8-10]. Sorption rates, and especially the normalized desorption rates that are developed in this paper, are a fruitful means of addressing adsorption-desorption hysteresis and reversibility issues.

One of the chief objections to the use of an equilibrium adsorption modeling approach to establish criteria for non-toxic metal concentrations in sediments has been the lack of evidence that adsorption and desorption can be appropriately treated as equilibrium processes [11,12]. The concern as to whether metal-sediment interactions can be treated as equilibrium systems derives from two types of data. First, incomplete reversal has been reported for the adsorption of Cd onto natural sediments [13], of Ni and Co on clay minerals [14] and of Cd, Ni, and Zn on goethite [15]. Second, major differences have been reported in the time required for cationic metals to equilibrate with marine suspended sediments, depending on whether the adsorbates were added in the laboratory or had been adsorbed as a result of the prior release of the metals into the environment [4,12]. For example, Piro *et al.* [16] reported that a pH-dependent equilibria between ionic, pH-labile- "complexed," and "particulate" (i.e., adsorbed) Zn in sea water. Decreasing the pH from 8 to 6 caused the desorption of adsorbed Zn and a further pH decrease to 2 released Zn from the pH-labile-complexed form. Release of stable Zn from the pH-labile-complexed form after electrolysis at pH 6 required about 16 and 24 hr to reach equilibria at pH 6 and 8, respectively. However, when electrolyzed at pH 8 to remove ionic Zn, there was no detectable desorption from the particulate or dissociation from the labile-complexed forms. Ionic [65]Zn added to natural seawater did not reach equilibria with the pH-labile-complexed form of stable Zn within a year. The pH-labile complexants in seawater are presumably derived from phytoplankton growth and decay and may be partially humified. Such effects have been ascribed by the Science Advisory Board [12] to an "aging process" that is said to occur during the interaction of metals and suspended marine particulates. According to the Science Advisory Board [12], "This implies that chemicals may be more bioavailable in toxicity studies with freshly spiked sediment than with sediments collected from the natural environment with the same chemical." If true, this could negate the application of laboratory data to the natural, or at least the marine, environment.

Concern about the degree to which equilibrium obtains in metal-sediment systems has led a number of investigators to check the reversibility of adsorption by sediments (e.g., [17-21]) and soils [22], but no consensus has evolved concerning the interpretation of differences in the apparent reversibility of adsorption. One reason for the lack of a common understanding of the reversibility of adsorption reactions is the wide differences in the time frames used by various investigators. One approach to resolving the conflicting reports concerning reversibility and the time required to reach equilibrium is examination of the rates of adsorption and desorption. Although virtually no attention has been given to such rates for sediments, it seems apparent that such rates are essential to resolve the ongoing controversy of not only "irreversible" adsorption [14] and desorption hysteresis [23], but the solids concentration effect [14,23,24] as well.

Sorption rates can be expected to be impacted by the factors that affect the adsorption capacity of sediments for cationic metals. These include pH, pretreatments such as oven- or freeze-drying [25], complexation of metals with organic and inorganic ligands, [18,26] and competition with other dissolved metals [27]. However, as these variables are only occasionally characterized when they are not being

explicitly investigated, these effects cannot generally be factored out of
rates derived from literature data.

2. MULTIPLE REACTIONS

Sediments typically contain multiple adsorbents (e.g., amorphic
and crystalline Fe oxides, poorly crystalline Mn oxides, particulate
organic carbon [POC], and carbonates [28]), each of which may have
multiple types or locations of adsorption sites. As a consequence, the
observed overall rate of adsorption onto or desorption from sediment is
a composite of multiple reactions. Both fast and slow reactions have
been observed for such nominally single-phase absorbents as oxides or
clay minerals [15,18,19,29-31]. Similarly, cationic metal adsorption
onto and desorption from sediment nearly always displays both a
relatively fast and a relatively slow reaction [8,12,21,32,33,34]. A
number of investigators have attributed the slow process to the diffusion
of metals into sediment particles [15,35,36] or of organic compounds
into sediment particles (e.g., [37]). Given that the fast and slow
processes are both likely to represent multiple reactions and that the
available data are inadequate to allow separation of possible multiple
reactions, the fast and the slow reactions are treated in this paper as
being single slow and fast processes. Process is generally used in
preference to reaction in this paper because the reaction cannot be
specified in the case of sediments. Treatment of adsorption and
desorption as diffusion controlled processes implies that the rate limiting
processes are diffusion controlled, not that all individual reactions are
diffusion limited.

3. DATA CALCULATION AND PLOTTING [1]

Appropriate graphical displays of adsorption data facilitates its
interpretation. Adsorption isotherms have traditionally been presented

---

[1] Abbreviations used in figure notes and text are as follows: ' = feet; % = salinity; A
= activity; ads = adsorption; aq = aqueous; AVS = acid volatile sulfide; BET =
Brunauer-Emmett-Teller surface area method; C = concentration of indicated (i.e.,
subscripted) item; CA = California; CEC = cation exchange capacity; cpm = counts
per minute; d wt = dry weight; des = desorption; det = detection; DOC = dissolved
organic carbon; E = east; f = final; i = initial; I = ionic strength; K_d = distribution
coefficient; L = liter; M = moles; m = meters; Me = metal; N = normalized; OAc =
acetate; POC = particulate organic carbon; SA = specific surface area; sed =
sediment; soln = solution; susp = suspended; $time^{1/2}$ = square root of time; T =
temperature; unrep = unreported; w/ = with; WA = Washington.

as a fractional or percent adsorption or as a distribution coefficient versus some variable; the distribution coefficient, K_d, differs importantly from fractional or percent adsorption in that it is normalized to sediment concentration. Although percent or fraction adsorbed (mass Me adsorbed/total mass of Me in system) is a convenient mode of presentation as pointed out by Benjamin and Leckie [38], it is not particularly informative inasmuch as the effects of sediment and metal concentrations, C_{sed} and C_{Me}, on adsorption density are exaggerated. The K_d is superior to fraction or percent adsorbed, in that the fraction adsorbed is normalized to the sediment concentration (L/g sed). An additional limitation of the fraction or percent adsorbed format is that its use has led investigators to emphasize the importance of pH and disregard the decrease in $C_{Me,aq}$ that occurs with adsorption as the pH is increased. For these reasons, the primary mode of presentation of adsorption data in this paper is as micromoles of metal per unit mass of sediment (μM Me/g Sed). When published experimental information was inadequate to allow calculation of the $C_{Me,ads}$, either the K_d or, if unavoidable, the fraction adsorbed or desorbed was used.

Linearization of adsorption data is desirable because it allows ready interpolation and extrapolation and facilitates identification of experimental values that are outliers. In view of suggestions that the slow adsorption process is diffusion limited, a few investigators have plotted their adsorption data versus the square root of time, time$^{1/2}$, to linearize the acquired data. For example, Davis et al. [34] linearized the early stage of Cd adsorption (fraction of amount adsorbed at equilibrium) onto calcite with a time$^{1/2}$ plot. Similarly, Harter [39] studied Ni adsorption by a soil and found that plotting $\Delta C_{Me,time=t}/C_{Me,i}$ versus time$^{1/2}$ linearized the data. Avotins [29] found that Hg adsorption on Fe(OH)$_3$(amorph) was linear on time$^{1/2}$ plots. Elkhatib and Hern [40] plotted the amount of K (mg K/kg) desorbed from a soil versus time$^{1/2}$ to linearize their data and obtained distinct curves for each initial concentration of adsorbed K. Bruemmer et al. [27] found that the time$^{1/2}$ function linearized the adsorption of Cd, Ni, and Zn onto goethite.

Although there is a tradition of plotting $C_{Me,aq}:C_{Me,ads}$ versus $C_{Me,aq}$ to approximately linearize adsorption data, e.g., Cu and Ni adsorption on soil [41], normalizing the $C_{Me,aq}$ to $C_{Me,ads}$ is illogical whereas normalizing $C_{Me,ads}$ to $C_{Me,aq}$ (e.g., [42,43]) serves to both provide a number which can be usefully compared among experiments and serves to remove one of the master variables (i.e., $C_{Me,aq}$), allowing a more meaningful graphical analysis of other effects such as pH. More frequently the log $C_{Me,ads}$ is plotted versus log C_{aq} to linearize adsorption or desorption at a single solids concentrations

[29,44-46].   Avotins [29] noticed that the fit of this relationship decreased at pH 11 where the adsorption rate was slower than at lower pH values.

For this study, published data were entered from tables into a spreadsheet. However, more frequently the experimental values were obtained from figures by digitizing to an ASCII (DOS) file, which was then imported into a personal computer spreadsheet.   Extensive calculations often had to be made to recover the data in the desired units. A uniform set of units is essential for comparing results between investigators.   Units of micromoles of metal adsorbed per gram sediment were selected.   For K_d values, the units of liter per gram are used to maintain consistency with the molar units for aqueous concentrations.   (No conversion was carried out for data from those occasional investigators who render the K_d dimensionless, e.g., [47]).

Because the published adsorption or desorption data for sediment rarely have enough experimental points or precision to allow meaningful treatment with rate equations, the overall fast and slow reactions were treated graphically.   The arithmetic plot of $C_{Me,ads}$ versus time$^{1/2}$ facilitates identification of the presence of fast and slow adsorption reactions, since this curve must start at or subsequent to (where mixing is inadequate and bulk diffusion is involved) zero time. Extrapolation of the slow reaction to zero time provides an estimate of the $C_{Me,ads}$ attributable to the fast reaction [33].   It was found that normalizing the $C_{Me,ads}$ to the associated $C_{Me,aq}$, or activity ($A_{Me,aq}$), allows ready comparison of different metal and sediment concentrations. Because the range in $C_{Me,ads}:C_{Me,aq}$ and $C_{Me,des}:C_{Me,aq}$ are often orders of magnitude, it is convenient to plot the logarithm of this ratio. Adsorption and desorption rates versus time were linearized via log-log plots. Since adsorption and desorption rates are strongly influenced by the amounts of metal already adsorbed or not yet desorbed, it is useful to normalize the adsorption rates, $R^N_{Me,ads}$ or $R^N_{Me,des}$ ($\Delta\mu$M Me/g Sed/hr), to the total amount of metal already sorbed or not yet desorbed at the end of successive time intervals to obtain a normalized rate, $R^N_{Me,ads}$ or $R^N_{Me,des}$ [($\Delta\mu$M Me/g Sed/hr)/(μM Me/g Sed)].   Because the apparent reaction rate may also range over orders of magnitude, it is convenient to plot the logarithm of $R^N_{Me,ads}$ or $R^N_{Me,des}$ versus log time. Visual perceptions of the scatter, i.e., closeness of fit of experimental points in graphically displayed data to a smooth curve, are greatly impacted by the scale at which the data are plotted. Therefore, to the extent feasible, the logarithmic plots presented in this paper generally cover five and sometimes three decades, but occasionally less, to facilitate comparisons among plots.

Data have been recovered from several individual papers, and the inclusion of details of the experimental procedures from those papers in the present paper would make it excessively long and would require the readers to frequently refer to the methods section while examining the figures. To circumvent these problems to a degree, salient aspects of the various experimental procedures, especially factors affecting the rate and extent of adsorption, are given in abbreviated form in figure notes.

4. RESULTS

The graphical data reduction procedure described above was used to evaluate the utility of time$^{1/2}$ plots and the comparability of adsorption and desorption rates across metals and investigators. Because of the quality of the data and auxiliary information available, the adsorption data of Fu [48], desorption data of Lion *et al.* [30], and desorption data of Kennedy *et al.* [49] were used in this comparison. Although this study is certainly not exhaustive, adsorption, desorption, and rate data from several other investigators are included in a subsequent section of this report. Adsorption data obtained in systems where bulk diffusion is a significant factor are presented later in a separate section of this report because such data are not susceptible to the data reduction procedures used for systems without bulk diffusion.

4.1 Fu [48]

Figure 1a is a typical metal adsorption curve except that $C_{Me,ads}$ has units of micromole Cd/gram sediment instead of fraction or percent and $C_{Me,ads}$ is plotted versus time$^{1/2}$ instead of time [21]. Regression lines using two and five points indicate the initial fast reaction and the subsequent slow adsorption reactions, respectively. The zero-time intercept is 2.24, which is 94% of the total amount adsorbed in the 24-hr experiment. Where the number of experimental points is so small, the decision as to which points to include in individual regression lines is clearly subjective.

Most investigators assume that a more-or-less linear curve with low slope on a percent adsorption versus time plot is sufficient evidence that a particular reaction time is adequate for equilibrium. However, it is important to note that the $C_{Cd,aq}$ is decreasing continuously as $C_{Cd,ads}$ increases, as shown in Figure 1b. Thus, at the beginning of the nominally straight line fitted to the last five experimental points, $C_{Cd,aq}$ had fallen from its original value by $\approx 33\%$. The effect on the $C_{Cd,ads}$ of the continuous decrease in $C_{Cd,aq}$ is partially removed by normalizing

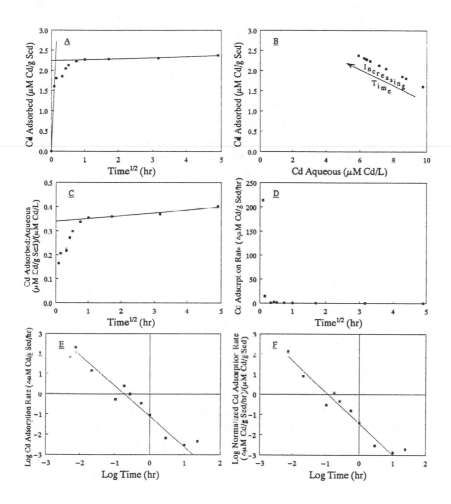

Figure 1.   *Adsorption of Cd on a Fluvial Sediment as a Function of Time (After Fu [48], Fig. 5-6). [Duwamish River, WA; CEC (w/NH₄) = 14.6 meq/100 g; NH₂OH·HCl extractable (mg/g) Fe = 8.49, Mn = 0.30; AVS = <13 ug S(II⁻)/g d wt sed; BET SA = 16.2 m²/g; pH (50 g/L, 0.1M NaNO₃, w/N₂ bubbling) = 5.8; Solution - 0.1 M NaNO₃, pH = 5.5, $C_{Cd,i}$ = 1.78×10⁻⁴ M, T = 25°C; C_{sed} = 50 g/L d wt; 0.45 μm filtration.]*

$C_{Cd,ads}$ to the $C_{Cd,aq}$; the slope of the slow reaction increased perceptibly (Figure 1c). Other data not shown here [50] indicate that the time to apparent equilibrium is dependent on the metal:sediment ratio.

The adsorption rate, $R_{Cd,ads}$, falls precipitously from 214 to 14.3 to 0.5 $\Delta\mu M$ Me/g Sed/hr over the initial 0.45-, 0.85-, and 5.04-min intervals (Figure 1d). These data are displayed in a more useful manner as log $R_{Cd,ads}$ versus the logarithim of time in Figure 1e. Since $C_{Cd,ads}$ would be expected to decrease with decreasing availability of sites, the $R_{Cd,ads}$ is divided by the concentration already adsorbed to obtain $R_{Cd,ads}^N$, which is also plotted against log time (Figure 1f). This normalized rate can be compared with similarly derived rates from other studies.

4.2   Lion *et al.* [30]

Lion *et al.* [29] obtained a typical sigmoidal curve of $C_{Me,ads}$ versus pH for Pb and Cd adsorption on oxidized estuary sediment (Figure 2a). Although it is not readily apparent from Figure 2a, the metal:sediment ratio for Pb is 400 times that of Cd ($C_{Pb,aq,i}$ is 100 times that of $C_{Cd,aq,i}$ and C_{Sed} is 1/4 that used for Cd). However, as can be seen in Figure 2b, normalizing the $C_{Me,ads}$ to associated $C_{Me,aq,i}$ yields $C_{Me,ads}:C_{Me,aq,i}$ values that are generally less than a factor of 10 higher for Pb than Cd.

In a separate experiment, Lion *et al.* [30] adsorbed Pb and Cd onto additional aliquots of the sediment at pH 7.1 and 7.6, respectively, for 24 hr and then desorbed these metals by lowering the respective pH values to 5.0 and 5.5. For both Pb and Cd, fast as well as slow desorption rates were demonstrated; about 40 times more Pb than Cd was desorbed (Figure 2c). The absolute desorption rate in this system for Pb during the 0.25- to 0.5-min interval is 3.2 μM/g Sed/hr, in contrast to 0.008 μM/g Sed/hr for Cd (Figure 2d). However, the $R_{Pb,des}$ is less than the $R_{Cd,des}$ because much more Pb than Cd was originally adsorbed (Figure 2e,f).

4.3   Kennedy *et al.* [49]

Kennedy *et al.* [49] carried out a stream-spiking (K, Sr, Cl) experiment to determine the rates of solute metal removal by bed sediment. They size-separated the <4,000-μm portion of the bed sediment from a small creek and placed porous bags of several size fractions of stream sediment on the bed surface just prior to stream spiking. They subsequently measured the desorption of Sr and K, by sequential batch extraction with ammonium acetate, from both the pre-

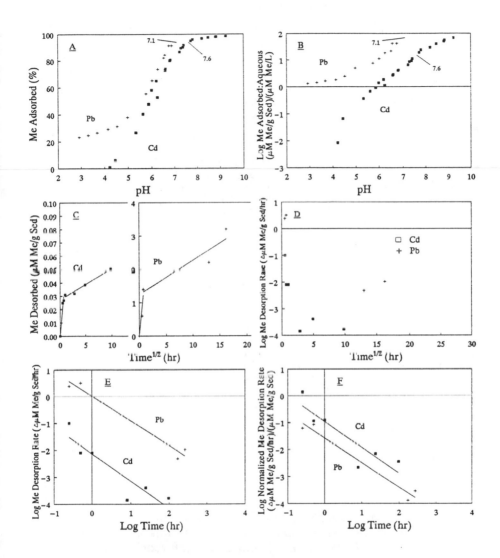

Figure 2.   *Adsorption of Cd and Pb Versus pH and Subsequent Desorption Versus Time by Oxidized Surficial Tidal-Flat Sediment (After Lion et al. [30], Fig. 1a,b and 4a,b). [South San Francisco Bay, CA; Sampled June 9 and Aug. 8, 1980; Storage - pH ≈ 8, 4°C, Continuous stirring; Sed - Sample B; Soln - 0.6 M NaNO₃; Cd - C_{sed} = 1 g/L, $C_{Cd,i}$ = 10^{-7} M; Pb - C_{sed} = 0.25 g/L, $C_{Pb,i}$ = 10^{-5} M; Des - Preceding ads - at pH to give 90% ads (7.1 for Pb, 7.6 for Cd), Time = 24 hr, des pH - Cd = 5.5 and Pb = 5.0.]*

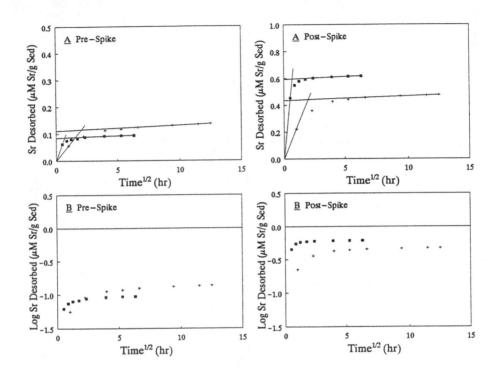

Figure 3.   *Desorption of Sr Versus Time$^{1/2}$ for Sized Fluvial Sediment (Collected Before and After Stream Spiking Experiment) with NH$_4$OAC Using Sequential Batch Extractions (Size in μm - ▨ = 210 to 250, + = 3,360 to 4,000; After Kennedy et al. [49], Figure 3 and personal communication). [Uvas Creek, Santa Clara Co., CA; Cobblestone bed; <2-μm sed. (3.2 to 7.8% of <4 mm material) = 22% illite, 37% chlorite, 41% smectite+vermiculite and mixed layer clays; Preceding adsorption - sed placed in porous polyester bags and equilibrated in stream during upstream K+Sr+Cl injection (pH 8.3-8.5, 13-15.6℃); Desorption - 20 ml 1 M NH$_4$OAc added to damp 5-g sample, shaken for variable time (0.5-72 hr), centrifuged, filtered through 0.45-μm membrane; Total amount adsorbed taken as accumulative amount desorbed; Plotted at midpoint of extraction time].*

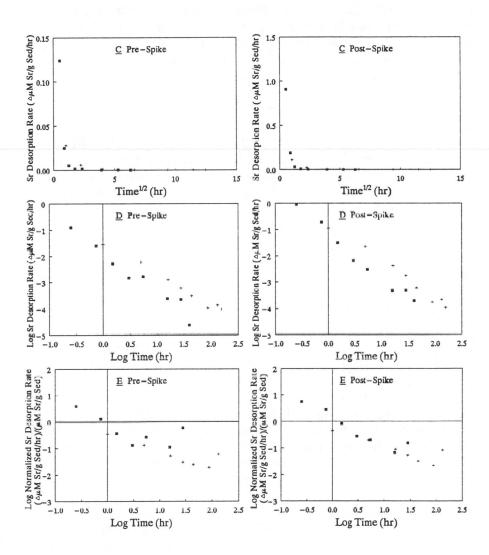

Figure 3.   *(Continued).*

spike and post-spike samples. Strontium desorption by the batch technique comprises both a fast and a slow reaction (Figure 3a). The quantity desorbed was 40 to 60% greater for post-spike than for the pre-spike samples (Figure 3a,b). The $R_{Sr,des}$ fell precipitously from the first to the second time interval. The $R_{Sr,des}$ was initially an order of magnitude greater for post-spike than pre-spike samples (Figure 3c,d), confirming the assumption inherent in the normalization step that desorption rates are highly dependent on the amount available for desorption. The $R_{Sr,des}$ approximated a linear log-log relationship for each size fraction, the smaller size fraction having the more rapid $R_{Sr,des}$. The normalized $R_{Sr,des}$ significantly reduces the difference in $R_{Sr,des}$ between size fractions and between the pre-spike and post-spike samples (Figure 3e). There is increased scatter of the points at the longest time intervals when the desorption rate becomes relatively small. A significant portion of the scatter in the $R_{Sr,des}^{N}$ values on the log-log plot may be a consequence of the wide range in extraction periods used (i.e., 1 to 156 hr).

4.4   Partially mixed systems with bulk diffusion

To include bulk diffusion as an explicit component of the investigation, some studies with marine sediment-water systems have involved bulk aqueous diffusion before adsorption ([6,47] and references therein). Duursma and Bosch [47], using the "thin-layer" technique[2], found that adsorption of Co and Zn continued for the 173-day duration of their experiment, although in the presence of penicillin adsorption reached apparent equilibrium in about half the total time of the experiment (Figure 4). The effect of penicillin may be at least partially an artifact of the thin-layer technique, wherein the metal adsorbed by suspended bacteria is attributed to the aqueous phase rather than to the sediment phase. There are too few short-time points to determine whether the fast adsorption reaction starts approximately at zero or if the fast-reaction extrapolates to a positive x-axis value. It appears that the thin-layer technique reduces the fast reaction, but the data are too limited to indicate whether the slow reaction is also slowed.

---

[2]In the "thin-layer technique" [47], 150 ml of 0.45-µm filtered seawater was taken to sterile Pyrex dishes and refrigerated at 4∞C overnight; then 1 ml of an isotope-containing solution was added and mixed, and an aliquot was taken for a control sample. "About 6" filters each containing ≈10 mg of Mediterranean fine sediment (located in a 1-cm round area in center of filter) was then placed in the bottom of the dish with a spoon. Gentle suction was applied to remove excess water. At approximate times, a filter was removed and placed on a suction apparatus and excess moisture was sucked off.

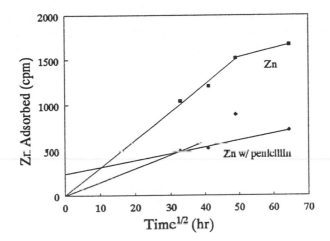

Figure 4. Adsorption of Zn by Marine Sediment Versus Time as Affected by 100,000 units/L of Penicillin (After Duursma and Bosch [47], Figure 29). [Mediterranean Sea, surficial, fine-grained, freshly sampled; Ads - "Thin-layer" technique, $C_{sed} = 10$ mg/filter, $C_{radioisotope} =$ unrep.]

Nyffeler *et al.* [6] added a single Cs spike (Figure 5a) and three successive Co and Zn spikes (Figure 5b,c) to a MANOP mockup chamber[3] (a 10-L cylindrical container with a ≈2-cm layer [900 cm²] of sediment over the bottom). Slow mixing of the overlying water was effected by a pump that circulated the water at a rate of 50 cm³/min. These chambers were designed to estimate rate constants for the diffusive flux of metals from the ocean floor. The decrease in $C_{Me,aq}$ measured near the top of the chamber was replotted as if it were adsorption data; the intervals between spike additions of Co and Zn were subtracted out so that the losses from solution following successive spike addition could be compared directly (Figure 5b,c). The data of Nyffeler *et al.* [6] appear to indicate that with each successive spike a slightly longer reaction time was required to reach the same adsorbed:aqueous ratio (Figure 5b,c). If this difference in the time axis intercept of the Co spike is subtracted from the times of spike 3, then the fractional adsorption curves for Co spikes 2 and 3 are superimposed (Figure 5d). Thus, the offset between spikes in Figure 5b and 5c may reflect a lack of precision in specifying the times of the spike additions. Extrapolation of the early time points on arithmetic plots of adsorption results in a time-axis intercept that indicates the passage of several hours before sufficient mixing and bulk diffusion had

[3]An ocean floor lander for carrying out *in situ* geochemical experiments on the ocean floor at depths to 5,000 m.

occurred in the 10-L chamber to allow a significant amount of adsorption. This delay was incorrectly attributed by Nyffeler *et al.* [6] to the "limiting film resistance".

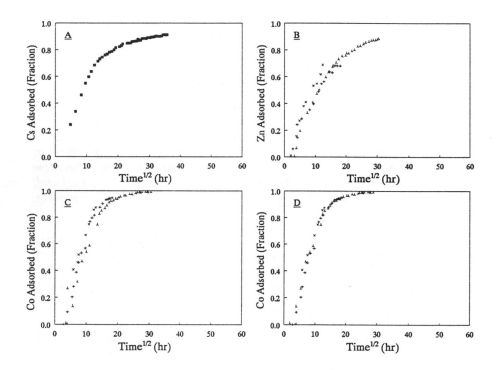

Figure 5.   Adsorption of Metals, onto Marine Sediments in MANOP Chamber Mockup System, Versus Time (<u>A</u> Single spike; <u>B</u>, <u>C</u>, and <u>D</u> multiple spikes of Co and Zn - x = 1st, + = 2nd, λ = 3rd; After Nyffeler et al. [6], Figures 2 and 3).   [San Clemente Basin, Pacific Ocean; Freshly collected sed; 2 cm sed over 900 cm²; ≈10 L freshly filtered seawater circulated at 50 cm³/min from top; Co and Zn radioactive spikes added at days 1, 18, and 39 but intervening periods subtracted out in <u>B-D</u>; <u>D</u> X-axis intercept subtracted from spikes 2 and 3; Co valence unrep.]

5.   THESES CONCERNING CATIONIC METAL ADSORPTION AND DESORPTION

On the basis of the graphical analyses of published data carried out in this study, the following theses have been developed.

1.   Adsorption or desorption of cationic metals from sediments comprises one or more fast and slow processes and the slow process is diffusion limited.

Although this conclusion has been reached by a number of investigators for individual studies, its generality has not been widely accepted. For example, no kinetic adsorption studies and no surface complexation constant studies have been found that consider the implications of the slow rate in their calculations. All time-dependent adsorption and desorption data analyzed in this study support the generality of fast and slow processes. The ubiquitous presence of fast and slow reactions is shown by Figures 1a, 2c, 3a, 4, 7, 8, 9, and 11.

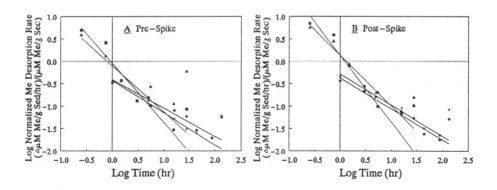

Figure 6.   Comparison of K and Sr Normalized Desorption Rates for Sized Sediment (Collected Before and After Stream Spiking Experiment) with NH₄OAC Using Sequential Batch Extractions (Size in μm - ⊠ *= 210 to 250 w/K,* ◆ *= 3,360 to 4,000 w/K,* Δ *= 210 to 250 w/Sr, x = 3,360 to 4,000 w/Sr; After Kennedy et al. [49], Figure 3 and personal communication). [Regression curve of minimum error (i.e., w/out last 1 or 2 values) drawn; See Figure 3 for other details.]*

Figures 6 and 10 are not relevant to this point, the data of Figure 5 are
confounded by bulk diffusion, and the results shown in Figure 12
appear strange (as if sufficient acidity was introduced with the
radionuclide spikes to inhibit adsorption for an initial period of time and
as if subsequently bacteria growth were competing with the sediment
for the radionuclides). Because these calculated adsorption rates are

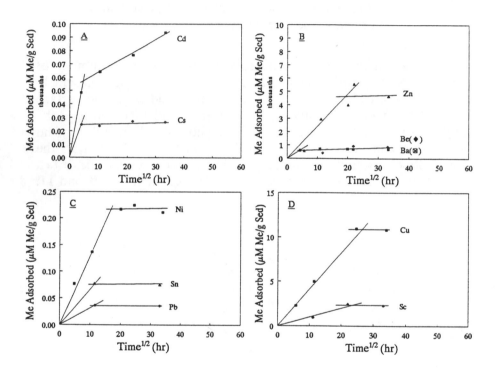

Figure 7. *Adsorption K_d Values for Marine Interfacial Sediment
Versus Time (After Balistrieri and Murray [51], Figure 1).
[MANOP Site H sed (≈1,000 km E of E Pacific Rise,
Guatemala Basin, ≈3,600 foot depth); Stored at 5°C; 60°C d
wt. basis; POC = 1.28%; C_{sed} = 1 mg/L (Be, Cu, Pb, Sc,
Sn, Zn) or 100 mg/L (Ba, Cd, Cs, Ni); Seawater pH =
7.82±0.04; C_{Cu} and C_{Ni}=3×10⁻⁸ M; $C_{radioisotopes}$: Ba =
4.44-44.4×10⁻¹¹, Be = 6.63-68.7×10⁻¹³, Cd = 2.7-
54.1×10⁻¹², Co = 6.41±0.56×10⁻¹⁴, Cs = 2.26-22.6×10⁻¹¹,
Fe = 5.46-32.7×10⁻¹¹, Pb = 2.32-6.81×10⁻¹¹, Mn =
<1.11×10⁻¹¹, Sc = 1.33-33.4×10⁻⁹, Sn = 3.08-25.7×10⁻¹¹,
Zn = 7.83-×10⁻¹²; Ads in 50 ml tubes, mixing unrep.]*

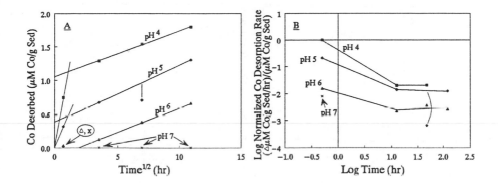

Figure 8. *Desorption of Co from Lake Sediment as a Function of Time at Four Nominal pH Values (After Yousef and Gloyna [52], Tables 4-8). [Lake Austin, TX, immediately upstream of Tom Miller dam; Composition - calcite+dolomite = 60%, clay fraction = 28%, quartz = 10%, POC = 2.6%; Dried to 103°C, ground to <200 mesh before cation saturation, redrying unrep; Ads - 7 days, C_{CoCl_2} = 10^{-6} M, C_{sed} = 0.5 g/L; Des - 7 days, presumably initial pH values, mechanically agitated; Centrifuged 10 min at 3,000·gravity; In Fig. a, $C_{Co,des,pH7}$ = 0, except for initial value.]*

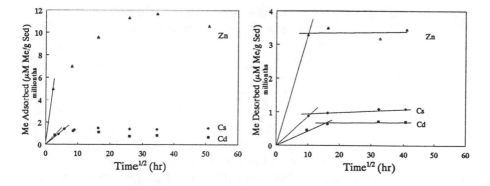

Figure 9. *Adsorption and Desorption of Metals on Estuarine and Marine Sediments Versus Time (After Nyffeler et al. [5], Figures 1a and 1b). [Narragansett Bay, Atlantic Ocean; Surficial sediment; C_{Sed} = 0.1 g/L (Adsorption) and 1.67 g/L (Desorption); Seawater - ‰ = 30, 0.4 μm filtered, pH 8.1; C_{spike} = 10^{-14} to 10^{-16} M, $C_{carrier}$ = 10^{-11} to 10^{-13} M, median value of 10^{-12} used in calculations; Stirred.]*

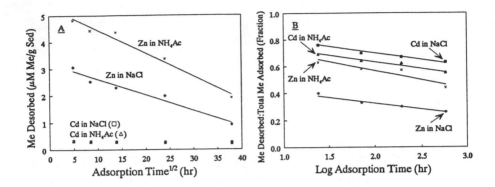

Figure 10.   *Desorption of Cd and Zn from River Sediment in 19‰*
NaCl and 1 N NH₄Ac Following Different Adsorption
Times (After Salomons [13], Figure 1, Table 1). [Rhine
River, Germany; Ads - presumably $C_{Cd,i}$ = 5 µM/L, $C_{Zn,i}$
= 50 µM/L, C_{sed} = 100 g/L; Des - Presumably C_{sed} = 100
g/L, presumably NH₄Ac pH ≈ 7, time unrep.]

functionally related to total number of adsorption sites, $C_{Me,aq}$, and pH
of the aqueous phase, at some adsorbate:adsorbent ratio only the "fast"
adsorption process will be observed [50].

Where sufficient experimental values are available, the fast process
is indicated by time$^{1/2}$ linearization to be diffusion controlled.
Frequently, however, only one or two experimental points are available
to indicate the possible slope of the fast process. Such a curve
represents a minimum value as experimental measurements taken at
some shorter times may have indicated a steeper slope of the fast
process. However, this is not a problem because the extent of the fast
process is obtained from the zero-time y-axis intercept of the slow
process. Analysis of single-phase oxide systems (data not shown)
indicate that experimental points falling between the fast and slow
reaction curves represent a period during which the rate of the fast
reaction decreases as the fast reaction sites fill, i.e., these points
represent the sum of the expiring fast reaction and the slow reaction.
The zero-time intercepts that are used to estimate the extent of adsorption
due to the fast process may be minimum values, given that it appears
likely that the regression slope of the slow reaction may be affected by
the length of the experiment. The intercept value for a particular
sediment is expected to vary depending on the concentrations of the
various adsorbents present and the degree of particle aggregation and
cementation.

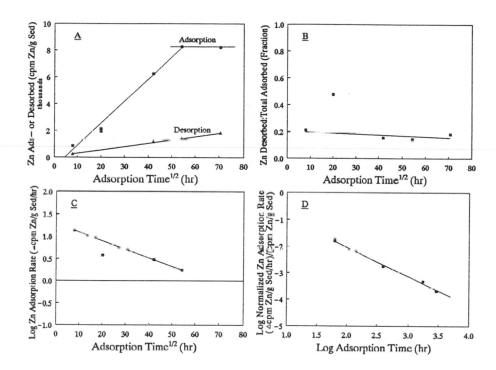

Figure 11.   *Zinc Adsorption and Desorption Versus Preceding*
Adsorption Time for a Pond Sediment (After Duursma
[53], Figure 3, from Ros Vicent et al. 1976, unpublished
manuscript). [Sediment type unrep., source unrep; Ads -
"Semi-natural" conditions in a pond; C_{sed}/water volume
unrep, $C_{Zn,i}$ unrep., water chemistry unrep.; Des - Soln
unrep (may have been 1 M acetic acid of pH 2.3); Des -
time presumably fixed but unrep. (p. 166), desorption
presumably done shortly after sampling, mixing unrep.,
400 hr outlier omitted from regression.

The plateauing of traditional adsorption curves is significantly impacted by the associated decrease in $C_{Me,aq}$, the extent of which is a function of $C_{Me,i}$, C_{sed}, and the adsorption capacity of the sediment. Thus, a reaction time that is adequate to reach a constant slow-reaction slope for one metal:sediment ratio may not be adequate for another. The effect of the continuous decrease in $C_{Cd,aq}$ can be partially removed and the comparability among experiments increased by normalizing the $C_{Me,ads}$ to the $C_{Me,aq}$ (e.g., Figure 2b). This ratio, with limits of ml/g, is the well known distribution coefficient.

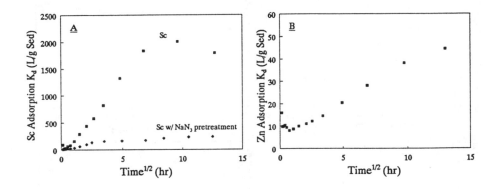

Figure 12. *Adsorption K_d for Suspended Sediment at Natural Concentration Versus Time (After Jannasch et al. [4], Figure 2). [Puget Sound, Pacific Ocean; $C_{sed} = 1$ mg/L but 1.49 mg/L for Sc w/NaN$_3$; Seawater - ‰ = 28, pH = 7.9, 12 °C, Carrier-free radioisotopes; Sc w/NaN$_3$ pretreatment in dark, others in light; Magnetic stirring; Mass balance of ^{45}Sc(III) and ^{65}Zn(II)>90%.]*

The frequency with which the slow reaction is linearized by plotting $C_{Me,ads}$ or $C_{Me,des}$ versus time$^{1/2}$ supports suggestions in the literature that the slow reaction is diffusion limited. It appears likely, as suggested by Jannasch *et al.* [4], that the fast and slow reactions operate sequentially, i.e., adsorption via the fast reaction must precede initiation of the slow reaction. It also appears likely that the fast reaction involves adsorption onto external adsorption sites, whereas the slow reaction represents adsorption onto sites reached only by aqueous-phase diffusion. This is in accord with the earlier conclusion of Anderson *et al.* [36] that during Ag adsorption diffusion-limited monovalent and divalent cation exchange occurred in the "intermediate" structural layers in poorly crystalline manganese oxides.

Given that adsorption generally consists of both fast and slow processes, it is to be expected that box type adsorption modeling will require at least two "black boxes" [8,54].

2.   Literature reports of fixation, hysteresis, and
irreversible adsorption are a consequence of
desorption commonly being slower than the
preceding adsorption.

That desorption is slower than adsorption was noted several years
ago for MnO_2 by Murray [55] and more recently for sediments by
Comans and van Dijk [19]. None of the data sets examined provide a
comparison of adsorption and desorption rates. The terms "fixation,"
"hysteresis," and "reversibility" are widely used in the sediment and
soils literature but are rarely defined. More recently, the term
"irreversible adsorption" has been used to describe the same phenomena
[14]. The term "fixation" implies that an aqueous phase cation has
partitioned to the surface by non-reversible adsorption (or by some other
process such as precipitation); however, the working definition is
commonly that the desorption (or dissolution) is incomplete within the
time frame of the desorption experiment. "Hysteresis" refers to the
commonly observed discrepancy between the amounts of a metal
adsorbed and desorbed in the same time frame and is commonly used in
reference to column experiments. "Reversibility" is typically used as a
qualitative term and frequently as an adjective indicating that some
portion of the adsorbed metal was reversibly desorbed, i.e., within a
time frame similar to that used for adsorption. Fixation, hysteresis, and
reversibility are all terms reflecting the tendency of metals to be
desorbed more slowly than they are adsorbed because of the smaller
driving force during desorption. Therefore, it seems much more
meaningful to quantify the subject phenomena by use of $C_{Me,ads}$
normalized adsorption and desorption rates, which also provide ready
comparability across experiments and investigators, recognizing that the
rates will be affected by system variables (e.g., pH, complexation,
temperature, competition). Importantly, rates can be used to estimate
the time required for a given degree of completion of these reactions.

As noted by van de Meent *et al.* [8], the large decrease in $R_{Me,des}$
with time makes it infeasible to define K_d in terms of single forward and
backward rate constants.

3.   The longer the time over which the preceding
adsorption occurs, the smaller the fraction of the
adsorbed metal that is desorbed in a <u>fixed time</u>.

Figure 10a indicates that the amount of Cd and Zn desorbed in a
fixed time decreases with increased adsorption time (the change in
amount adsorbed with increasing adsorption time was not reported by
Salomons [13]). That is, adsorption reversibility is decreased as the
time interval between adsorption and desorption is increased. In

contrast, Figure 11a shows that amount desorbed increased as the amount of Zn adsorbed increased with time although the amount desorbed as a fraction of the total amount adsorbed remained nearly constant (Figure 11b). Insufficient auxiliary information is available to reconcile the difference in these two experiments, but the difference in results may be a result of different metal:sediment ratios or desorption times. The adsorption curve in Figure 11a intercepts the x-axis at 20.8 hr (time$^{1/2}$ [hr] = 4.56). This may approximate the mixing time of the ratioactive spike added to the pond.

Major differences in the extent of metal adsorption and desorption have been reported for K_d values of radionuclides determined in the laboratory and those calculated from ocean data [4,12]. The Science Advisory Board [12] attributes the difference in laboratory and field K_d values to an "aging process". Their contention that "The scientific literature clearly indicates that there are major differences in the adsorption characteristics of sediment-bound chemicals depending on whether the chemical is spiked onto the sediment or occurred as result of a release into the natural environment" (p. 11) apparently derives from disregarding the rates of these reactions. The reported discrepancy between laboratory and oceanic K_d values [4] appears to stem from disregarding the large differences in equilibration time between the laboratory and field data. The normalized desorption rates provide a means of estimating times required to approach equilibria for marine suspended sediments.

4.   Normalization of adsorption and desorption rates to the concentration of metal already adsorbed or not yet desorbed significantly reduces the differences among metals, between adsorption and desorption, among sediment size fractions, and among sediments.

Absolute adsorption rates vary by orders of magnitude as previously reported by van de Meent et al. [8]. These large differences occur as a result of differences in $C_{Me,ads}$ and $C_{Me,aq}$. This effect of $C_{Me,ads}$ is removed by normalization of adsorption and desorption rates to the total amount adsorbed by the end of the time interval.[4] This is nicely illustrated in Figure 3, where the $C_{Sr,des}$ is nearly 10 times greater for the post- than for the pre-spike sample but the $R^N_{Sr,ads}$ values coincide until near the end of the extraction series. As shown in Figure 2e and f,

---

[4]The accuracy of the log-log regression equations would presumably be increased by calculating and using the $C_{Me,ads}$ at the midpoint of the time interval and plotting the rates at the midpoint time interval. However, in view of the limited precision of the available data, this refinement did not appear generally worthwhile.

the absolute desorption rate for Pb is about two orders of magnitude greater than that of Cd, as a consequence of the much greater amount of Pb adsorbed from a higher initial metal concentration and lower sediment concentration. The normalized rates are much more similar, the $R_{Cd,des}^N$ being greater than $R_{Pb,ads}^N$, which would be expected given the higher adsorption energy of Pb. Significantly, the desorption rates of monovalent and divalent cations, $R_{K,ads}^N$ and $R_{Sr,ads}^N$, cover a similar range (Figure 6). The values of $R_{Pb,des}^N$ and $R_{Cd,des}^N$ shown in Figure 2 are similar to the $R_{Cd,ads}^N$ values displayed in Figure 1, although the desorption and adsorption experiments were carried out on different estuarine sediments. Furthermore, comparison of Figure 3d and e indicates that normalization largely removes grain-size effect in the data of Kennedy et al. [49].

The approach used here of calculating reaction rates for each time interval avoids the necessity to estimate a rate constant for adsorbent transfer between fast and slow adsorption compartments and the relative amounts of the two compartments.

5.   Adsorption and desorption rates are approximately log linear when the log rate is plotted versus log time.

Nearly all adsorption and desorption data graphed in this study from well-mixed systems with a given metal:sediment ratio yield permissibly linear curves when the adsorbed concentration is plotted versus time$^{1/2}$ (Figures 1e, 2e, 3d, 8b, and 11c). However, linearity is lost as the slow reaction approaches completion, as can be seen in Figures 3d, 3e, 6, and 8b.

The approximately linear log-log relationship between adsorption or desorption rate and time makes it possible to estimate the time to any given degree of reaction completion.

6.   "Equilibrium" adsorption and desorption are most usefully conceptualized as a rate rather than an absolute quantity.

Oxidized sediments appear to differ greatly in the time required to reach apparent completion of the fast process ($\approx$1 to 600 hr). The longer times appear to be associated with the use of tracer-level metal concentrations. Although the time to equilibrium is uncertain in most instances because the number of experimental points and the precision of the data is generally insufficient to reliably establish that equilibrium has been reached. A further complication is that the numerous variables [49] and the inherent chemical instability of sediment-water systems that have been perturbed affect adsorption. Thus, it is desirable to determine rate constants over a period of time and calculate the times required to reach any desired degree of reaction completion. The uncertainties in the analyses of low-levels of dissolved constituents or small changes in $C_{Me,ads}$ preclude the reliable determination of the amount of adsorbed metal as the reaction approaches completion. A further important factor affecting time to reach desorption equilibria is the duration of the preceding adsorption reaction (Figures 10 and 11). Thus, the soundest approach is to carry out adsorption or desorption measurements, calculate normalized adsorption or desorption rates, and calculate the time to a given degree of completion of the adsorption process. Intrinsic adsorption constants for use in surface complexation models of adsorption should be calculated from data collected with a reaction time that represents completion of the fast adsorption reaction. Environmental modeling of metals will clearly require combined thermodynamic and kinetic data.

6.   SUMMARY

Normalization of the adsorption or desorption rate, $R_{Me,ads}$ or $R_{Me,des}$ ($\Delta\mu M$ Me/g Sed/hr), to the quantity of that metal already adsorbed or not yet desorbed, $R_{Me,ads}^{N}$ [($\Delta\mu M$ Me/g Sed/hr)/(μM Me/g Sed)], greatly reduces the apparent differences in rates. Normalized rates have been calculated for Cd, K, Pb, and Sr metals, 210 to 250 and 3,360 to 4,000 μm size fractions, four sediments (one marine, two estuarine, and one fresh water), and one adsorption plus two desorption experiments. Although the data base is very limited, it is significant that the resultant normalized rates are of similar magnitude. If the first and last one or two points of adsorption and desorption, respectively, are disregarded, the values calculated from the data of Fu [48] and Lion *et al.* [30] fall within the same range. Similar normalized rates calculated from Kennedy *et al.* [49] occur at times that are approximately ten times longer, perhaps because of the relatively long individual extractions. Thus, the normalized rates are, to an extent, system-dependent. These normalized rates may be of value in dealing with environmental and policy problems, such as establishing sediment criteria for metals and biological availability of metals.

Acknowledgments: It is a pleasure to acknowledge the skillful and imaginative data reduction and plotting efforts of S.E. Faulk; the assistance of R.J. Avanzino and L.S. Balistrieri in providing either underlying or auxiliary data; technical review of V.C. Kennedy and J.S. Young, the insightful editorial review of L.K. Grove; and the library assistance of J.E. Madison. The assistance of B.L. Coffman and S. York in the development of the data reduction procedures is appreciated.

REFERENCES

1. Jenne, E.A., D.M. Di Toro, H.E. Allen and C.S. Zarba, "An Activity-Based Model for Developing Sediment Criteria for Metals," in *Proc. The International Conference on Chemicals in the Environment*, J.N. Lester, R. Perry, and R M. Sterritt, Eds., pp. 560-568 Selper Ltd., (London, 1986).

2. Shay, D., "Deriving Sediment Quality Criteria," *Environ. Sci. Tech.* 22:1256-1261 (1988).

3. van der Kooij, D., van de Meent, C.J. van Leeuwen and W.A. Bruggeman, "Deriving Quality Criteria for Water and Sediment from the Results of Aquatic Toxicity Tests and Product Standards: Application of the Equilibrium Partitioning Method," *Water Res.* 25: 697-705 (1991).

4. Jannasch, H.W., B.D. Honeyman, L.S. Balistrieri and J.W. Murray, "Kinetics of Trace Element Uptake by Marine Particles," *Geochim. Cosmochim. Acta* 52:567-577 (1988).

5. Nyffeler, U.P., Y.-H. Li and P.G. Santschi, "A Kinetic Approach to Describe Trace-Element Distribution Between Particles and Solution in Natural Aquatic Systems," *Geochim. Cosmochim. Acta* 48:1513-1522 (1984).

6. Nyffeler, U.P., P.H. Santschi and Y.-H. Li, "The Relevance of Scavenging Kinetics to Modeling of Sediment-Water Interactions in Natural Waters," *Limnol. Oceanog.* 31:277-292 (1986).

7. Santschi, P., P. Höhener, G. Benoit and M.B.-t. Brink, "Chemical Processes at the Sediment-Water Interface," *Marine Chem.* 30:269-315 (1990).

8. van de Meent, D., H.A. den Hollander and J.H. Verboom, "Sorption Kinetics of Micropollutants from Suspended Particles: Experimental Observations and Modeling," in *Proceedings of the 6th European Symposium of Organic Micropollutants in the Aquatic Environment*, G. Angeletti and A. Bjørseth, Eds., pp. 50-60 Kluwer, Dordrecht, (The Netherlands, 1991).

9. Oliver, B.G., "Biouptake of Chlorinated Hydrocarbons from Laboratory-Spiked and Field Sediments by Oligochaete Worms," *Environ. Sci. Tech.* 21:785-790 (1987).

10. Landrum, P.F. and J.A. Robbins, "Bioavailability of Sediment-associated Contaminants to Benthic Invertebrates" in *Sediments: Chemistry and Toxicity of In-place Pollutants*, R. Baudo, J.P. Giesy and H. Muntau, Eds., pp. 237-263. Lewis Publishers, (Chelsea, MI 1990).

11. Tessier, A. and P.G.C. Campbell, "Partitioning of Trace Metals in Sediments," in *Metal Speciation: Theory, Analysis and Applications*, J.R. Kramer and H.E. Allen, Eds., pp. 183-199 Lewis Publishers, (Chelsea, MI 1988).

12. Science Advisory Board. *Evaluation of the Equilibrium Partitioning (EqP) Approach for Assessing Sediment Quality*, Report of the Sediment Criteria Subcommittee of the Ecological Processes and Effects Committee (Washington, D.C.: U.S. Environmental Protection Agency, EPA-SAB-EPEC-90-006, 1990).

13. Salomons, W., "Adsorption Processes and Hydrodynamic Conditions in Estuaries," *Environ. Tech. Lett.* 1:356-365 (1980).

14. Di Toro, D.M., J.D. Mahony, P.R. Kirchgraber, A.L. O'Byrne, L.R. Pasquale and D.C. Piccirilli, "Effects of Nonreversibility, Particle Concentration, and Ionic Strength on Heavy Metal Sorption," *Environ. Sci. Technol.* 20:55-61 (1986).

15. Tiller, K.G., J. Gerth and G. Brümmer, "The Sorption of Cd, Zn and Ni by Soil Clay Fractions: Procedures for Partition of Bound Forms and Their Interpretation," *Geoderma* 34:1-16 (1979).

16. Piro, A., M. Bernhard, M. Branica and M. Verzi, "Incomplete Exchange Between Radioactive Ionic Zinc and Stable Natural Zinc in Seawater," in *Proc. Symposium Interaction of Radioactive Contaminants with the Constituents of the Marine Environment*, pp. 29-45 IAEA/SM-158/2 (1973).

17. Davis, J.A., "Adsorption of Trace Metals and Complexing Ligands at the Oxide/Water Interface," *Ph.D. Thesis*, Stanford University, Stanford, CA (1977).

18. Comans, R.N.J., "Adsorption, Desorption and Isotopic Exchange of Cadmium on Illite: Evidence for Complete Reversibility," *Water Res.* 21:1573-1576 (1987).

19. Comans, R.N.J., and C.P.J. van Dijk. "Role of Complexation Processes in Cadmium Mobilization During Estuarine Mixing," *Nature* 336:151-154 (1988).

20. Middelburg, J.J. and R.N.J. Comans, "Sorption of Cadmium on Hydroxyapatite," *Chem. Geol.* 90:45-53 (1991).

21. Paalman, M.A.A., C.H. van der Weijden and J.P.G. Loch, "Sorption of Cadmium on Suspended Matter Under Estuarine Conditions; Competition and Complexation with Major Sea Water Ions," *Water Air Soil Pollut.* 73:49-60 (1994).

22. Christensen, T.H., "Cadmium Soil Sorption at Low Concentrations: II. Reversibility, Effect of Changes in Solute Composition, and Effect of Soil Aging," *Water Air Soil Pollut.* 21:115-125 (1984).

23. Schrap, S.M. and A. Opperhuizen, "On the Contradictions Between Experimental Sorption Data and the Sorption Partitioning Model," *Chemosphere* 24:1259-1282 (1992).

24. McKinley, J.P. and E.A. Jenne, "An Experimental Investigation and Review of the 'Solids Concentration' Effect in Adsorption Studies," *Environ. Sci. Technol.* 25:2082-2087 (1991).

25. Thomson, E.A., S.N. Luoma, D.J. Cain and C. Johansson, "The Effect of Sample Storage on the Extraction of Cu, Zn, Fe, Mn and Organic Material from Oxidized Estuarine Sediments," *Environ. Sci. Technol.* 20:836-840 (1980).

26. van der Weijden, C.H., M.J.H.L. Arnoldus and C.J. Meurs, "Desorption of Metals from Suspended Material in the Rhine Estuary," *Netherlands J. Sea Res.* 11:130-145 (1977).

27. Bruemmer, G.W., J. Gerth and K.G. Tiller, "Reaction Kinetics of the Adsorption and Desorption of Nickel, Zinc, and Cadmium by Goethite: I. Adsorption and Diffusion of Metals," *J. Soil Sci.* 39:37-52 (1988).

28. Jenne, E.A. and E.A. Crecelius, "Determination of Sorbed Metals, Amorphic Fe, Oxidic Mn, and Reactive Particulate Organic Carbon in Sediments and Soils," in *Proc. 3rd International Conference on Environmental Contamination*, A.A. Orio, Ed., pp. 88-93 Selper Ltd., (London, 1988).

29. Avotins, P., "Adsorption and Coprecipitation Studies of Mercury on Hydrous Iron Oxide," *Ph.D. Thesis*, Stanford University, Stanford, CA (1975).

30. Lion, L.W., R.S. Altmann and J.O. Leckie, "Trace-Metal Adsorption Characteristics of Estuarine Particulate Matter: Evaluation of Contributions of Fe/Mn Oxide and Organic Surface Coatings," *Environ. Sci. Technol.* 16:660-676 (1982).

31. Wehrli, B., S. Ibric and W. Stumm, "Adsorption Kinetics of Vanadyl (IV) and Chromium (III) to Aluminum Oxide: Evidence for a Two-Step Mechanism," *Coll. Surf.* 51:77-88 (1990).

32. Morel, F.M.M., *Principles of Aquatic Chemistry* John Wiley and Sons, (New York, 1983).

33. van de Meent, D., H.A. den Hollander and J.H. Verboom, "Release Kinetics of Metals and PAH from Natural Sediment Particles," Poster presentation at the *5th IASWS Symposium on Interactions Between Sediment and Water* (Uppsala, Sweden, 1990).

34. Davis, J.A., C.C. Fuller and A.D. Cook, "A Model for Trace Metal Sorption Processes at the Calcite Surface: Adsorption of Cd and Subsequent Solid Solution Formation," *Geochim. Cosmochim. Acta* 51:1477-1490 (1987).

35. Duursma, E.K. and M.G. Gross, "Marine Sediments and Radioactivity," in *Proc. Radioactivity in Marine Environment*, Panel on Radioactivity in the Marine Environment of the Committee on Oceanography, National Research Council, pp. 147-160. U.S. National Academy of Science, (Washington, DC 1971).

36. Anderson, B.J., E.A. Jenne and T.T. Chao, "The Sorption of Silver by Poorly Crystallized Manganese Oxides," *Geochim. Cosmochim. Acta.* 37:611-622 (1973).

37. Karickhoff, S. and K. Morris, "Sorption Dynamics of Hydrophobic Pollutants in Sediment Suspensions," *Environ. Toxicol. Chem.* 4:469-479 (1985).

38. Benjamin, M.M. and J.O. Leckie, "Effects of Complexation by Cl, SO_4, and S_2O_3 on Adsorption Behavior of Cd on Oxide Surfaces," *Environ. Sci. Tech.* 16:162-170 (1982).

39. Harter, R.D., "Kinetics of Sorption/Desorption Processes in Soil," in *Rates of Soil Chemical Processes*, D.L. Sparks and D.L. Suarez, Eds., pp. 135-149. Soil Sci. Soc. Am., (Madison, 1991).

40. Elkhatib, E.A., and J.L. Hern, "Kinetics of Potassium Desorption from Appalachian Soils," *Soil Sci.* 145:11-19 (1988).

41. Harter, R.D. and R.G. Lehmann, "Use of Kinetics for the Study of Exchange Reactions in Soils," *Soil Sci. Soc. Am. J.* 47:666-669 (1983).

42. Benjamin, M.M. and J.O. Leckie, "Adsorption of Metals at Oxide Interfaces: Effects of the Concentrations of Adsorbate and Competing Metals," in *Contaminants and Sediments* (vol. 2), R.A. Baker, Ed., pp. 305-322 Ann Arbor Science, (Ann Arbor, 1980).

43. Benjamin, M.M. and J.O. Leckie, "Competitive Adsorption of Cd, Cu, Zn, and Pb on Amorphous Iron Oxyhydroxide," *J. Coll. Inter. Sci.* 83:410-419 (1981).

44. MacNaughton, M.G., "Adsorption of Mercury(II) at the Solid-Water Interface," *Ph.D. Thesis*, Stanford University, Stanford, CA (1973).

45. Benjamin, M.M. and J.O. Leckie, "Multiple-Site Adsorption of Cd, Cu, Zn, and Pb on Amorphous Iron Oxyhydroxide," *J. Coll. Inter. Sci.* 79:209-221 (1981).

46. Sharpley, A.N., "Kinetics of Sulfate Desorption from Soil," *Soil Sci. Soc. Am. J.* 54:1571-1575 (1990).

47. Duursma, E.K. and C.J. Bosch, "Theoretical, Experimental and Field Studies Concerning Diffusion of Radioisotopes in Sediments and Suspended Particles of the Sea. Part B: Methods and Experiments," *Netherlands J. Sea Res.* 4:395-469 (1970).

48. Fu, G., "Sorption of Metals to Oxic Natural Sediments," *Ph.D. Thesis*, Drexel University, Philadelphia, PA (1990).

49. Kennedy, V.C., A.P. Jackman, S.M. Zand, G.W. Zellweger and R.J. Avanzino, "Transport and Concentration Controls for Chloride, Strontium, Potassium and Lead in Uvas Creek, A Small Cobble-Bed Stream in Santa Clara County, California, U.S.A, 1. Conceptual Model," *J. Hydrol.* 75:67-110 (1984).

50. Jenne, E.A., "Metal Adsorption onto and Desorption from Sediments: 2. Artifact Effects," *Mar. Freshwater Res.* 64:1-18 (1995).

51. Balistrieri, L.S. and J.W. Murray, "Marine Scavenging: Trace Metal Adsorption by Interfacial Sediment from MANOP Site H, *Geochim. Cosmochim. Acta* 48:921-923 (1984).

52. Yousef, Y.A. and E.F. Gloyna, *Radioactivity Transport in Water - The Transport of Co^{58} in an Aqueous Environment*, Technical Report 7 (Austin, TX University of Texas, Environmental Health Engineering Research Laboratory, 1964).

53. Duursma, E.K., "Radioactive Tracers in Estuarine Chemical Studies," in *Estuarine Chemistry*, J.D. Burton and P.S. Liss, Eds., pp. 159-183 Academic Press, (New York, 1976).

54. Comans, R.N.J. and D.E. Hockley, "Kinetics of Cesium Sorption in Illite," *Geochim. Cosmochim. Acta* 56:1157-1164 (1992).

55. Murray, J.W., "The Interaction of Metal Ions at the Manganese Dioxide-Solution Interface," *Geochim. Cosmochim. Acta* 39:505-519 (1975).

5

METAL AND SILICATE SORPTION AND SUBSEQUENT MINERAL FORMATION ON BACTERIAL SURFACES: SUBSURFACE IMPLICATIONS

Susanne Schultze-Lam[1], Matilde Urrutia-Mera[2] and Terry J. Beveridge[2]
[1]Department of Geology
University of Toronto
Toronto, Ontario, Canada M5S 3B1
[2]Department of Microbiology
College of Biological Sciences
University of Guelph
Guelph, Ontario, Canada N1G 2W1

1.   INTRODUCTION

Gram-positive and gram-negative bacterial surfaces are good interfaces for the sorption of dilute environmental metal cations. This is due mainly to the anionic properties of the peptidoglycan and the secondary polymers which make up the gram-positive surface, and to the phosphoryl groups of the lipopolysaccharide which stud the gram-negative outer membrane. Once metal ions have interacted with the electronegative sites on these surface macromolecules, they nucleate the formation of fine-grained minerals by incorporating common anions from their surroundings, encouraging further metal complexation. Accordingly, hydroxide/oxide, carbonate, sulfate/sulfide, and phosphate minerals are common and can be in transition from amorphous phases to crystalline phases depending on the mineral development stage. Crystallinity usually correlates with increasing depth within the subsurface environment. In this review, evidence from numerous field and laboratory studies of bacterial involvement in mineral nucleation and formation is presented and discussed.

2.   BACTERIAL STRUCTURE AND CELL SURFACE CHEMISTRY

Like all other groups of living organisms, bacteria come in a vast range of shapes and sizes. Since bacteria are very small it is necessary to rely on microscopical techniques to discern individual cells and their components. Most bacteria range in cell diameter from 0.8 to 1.5 µm and 1.5 to 6 µm in length and are thus comparable in size to the clay fraction (< 2 µm) or even fine silt (2 to 50 µm, USDA classification) in soils and sediments. It is well known that clay-size particles are very important in soils and sediments due to their reactivity and physicochemical properties. Similarly, bacteria are also very reactive and influential entities in these environments.

Despite the great structural diversity that bacteria exhibit, this large, heterogeneous group of organisms can be roughly divided into two major groups, based on their response to a staining regimen developed for light microscopy by Christian Gram in 1884 [1]. Dried smears of cells are exposed to four different chemical agents (crystal violet, iodine, ethanol and safranin) in sequence. Cells that are able to retain the crystal violet-iodine complex despite "decolorization" by ethanol stain deep purple and are referred to as gram-positive. Alternatively, if the cells lose the crystal violet-iodine complex they become colorless and can be counterstained with safranin to become red; these are gram-negative. Davies and coworkers [2] developed a heavy metal analogue to replace the iodine which revealed the actual staining mechanism of bacteria representative of each of the two types by electron microscopy [3]. Gram-positive cells had a thick amorphous wall layer exterior to their plasma membrane while those which stained gram-negative had two lipid bilayer membranes, the inner (plasma) membrane and the outer membrane. Between these two layers was a thin, darkly staining layer of peptidoglycan. The term "wall" is used to refer to the layers exterior to the plasma membrane of both gram-positive and gram-negative bacteria. These wall types are described in greater detail below.

2.1   Gram-positive bacteria

In the electron microscopy the cell wall of most gram-positive bacteria such as *Bacillus subtilis* appears as an amorphous layer approximately 20 to 25 nm thick [4] (Figure 1a). When cells are prepared by freeze-substitution, a method in which the rapid freezing of the specimen reduces the incidence of preparation artifacts, the walls of *B. subtilis* and other gram-positive bacteria appear as a fibrous, brushlike layer [5]. Peptidoglycan is a major constituent of the wall (Figure 1b) and is composed of strands that form a linear backbone of chains up to 50 dimers (of N-acetylglucosamine and N-acetylmuramic acid) in length [6]. The peptide stems attached to the muramic acid residues are rich in carboxyl groups and provide highly reactive sites for metal sorption [7]. They also allow cross-linking to occur between adjacent glycan chains (Figure 2). Depending on the bacterium and its peptidoglycan chemotype, the cross-links may be direct or there may be interchain peptides present. The end result is a highly resilient structure that completely surrounds the protoplast and is peripheral to its enclosing plasma membrane [4]. The peptidoglycan layer is so resistant to degradation that, as long as its constituent autolysins (degradative enzymes) are denatured, it can persist in the environment long after the cell dies [8].

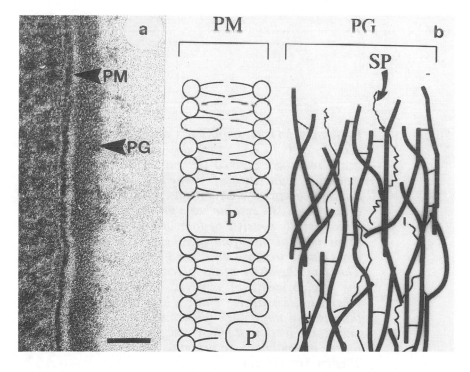

Figure 1.   *The Gram-positive Bacterial Cell Wall. When prepared by a standard embedding method, thin-sectioned, and viewed by TEM (transmission electron microscopy; A) the wall appears to consist of three layers: the plasma membrane (PM), the peptidoglycan layer (PG) and an electron-transparent region between them. Bar = 50 nm. These layers are shown in a schematic diagram (B).*

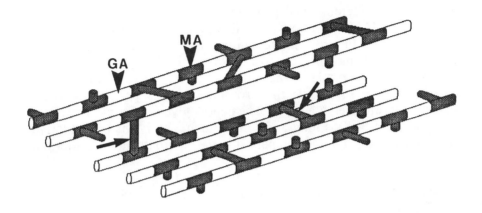

Figure 2.   *Peptidoglycan is an Important Constituent of all Bacterial Walls and Can Provide Binding Sites for Environmentally-derived Cations. The backbone of the molecule consists of repeated dimers of N-acetylglucosamine (GA) and N-acetylmuramic acid (MA). Each MA residue has a peptide stem attached to it which usually consists of four or fewer amino acids. A proportion of the four-membered peptide stems are cross-linked to those on other glycan strands (arrows). The stereochemistry of the glycan backbone is such that the peptide stems of adjacent dimers are rotated 90° relative to each other. Thus, when cross-linked, a three-dimensional meshwork is formed, which encloses the entire cell, lending it shape and protection from rupture due to internal osmotic pressure. (Drawing courtesy of Braden Beveridge.)*

Usually one of several types of secondary polymers are joined to and intermeshed with the peptidoglycan strands. The most frequently encountered are teichoic acids (polyglycerol phosphates or polyribitol phosphates with residues linked by phosphodiester linkages) or teichuronic acids (uronic acid-based polymers). These are bonded to a variable number of muramic acid residues and often make up 50 to 60% (dry weight) of the cell wall in bacillae [7]. In one of the most thoroughly studied gram-positive bacteria, *B. subtilis*, the type of secondary polymer present is influenced by the amount of magnesium and phosphate present in the culture medium. Cells grown in phosphate-limited medium possess only teichuronic acids, but will form teichoic acids exclusively when grown with sufficient phosphate and magnesium [9]. Other bacteria, such as *B. licheniformis,* possess both teichoic and teichuronic acids in their walls at the same time regardless of growth conditions [10,11].

Although other chemical constituents may be present in the gram-positive cell wall, the peptidoglycan and teichoic (or teichuronic acids) are the most important entities with respect to the ability of the wall to sorb metal ions. The presence of carboxyl and phosphoryl groups imparts an overall electronegative charge density which allows the cells to be potent scavengers of cations [11,12].

2.2 Gram-negative bacteria

In contrast, the walls of gram-negative bacteria appear as multilayered structures in the electron microscope (Figure 3a). External to the plasma membrane is the peptidoglycan layer which is overlaid by another bilayered structure, the outer membrane (Figure 3b). It is this membrane which first encounters the external environment and consequently is most important in considerations of metal binding by gram-negative bacteria.

The outer membrane is a bilayered structure that has an asymmetric lipid distribution; phospholipids are mostly limited to the inner leaflet [13]. The outer leaflet of the outer membrane contains the uniquely prokaryotic molecule, lipopolysaccharide. This molecule possesses three distinctly different chemical regions. The innermost is the "lipid A" region and is the most highly conserved part of the molecule. In *Salmonella typhimurium* this region consists of a dimer of phosphorylated N-acetylglucosamine residues to which are bound a number of fatty acyl chains. Attached to the lipid A moiety is a "core" which consists of a combination of unique sugars, such as 2-keto-3-deoxyoctonate and one of several types of hexoses, which are usually negatively charged due to the presence of phosphoryl or carboxyl groups. The exact type of sugars present in the core varies with bacterial species and strain. The most distal (and most variable) region of the LPS molecule is the "O-antigen" region, so called because it forms the basis of a taxonomically important serotyping scheme. The O-antigen consists of repeating units of specific sugar substituents which confer a hydrophilic surface to the cell [14].

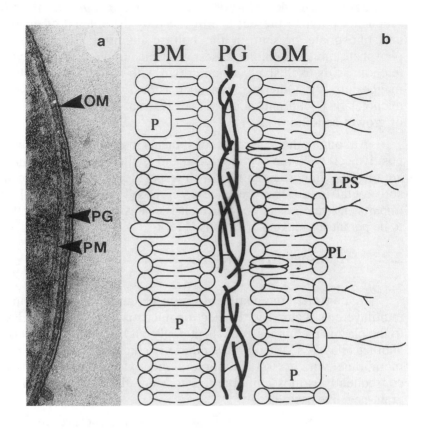

Figure 3.　　　*Gram-negative Bacterial Cell Wall Structure. TEM images (A) of thin-sectioned cells show a peptidoglycan layer (PG) flanked by bilayered membranes on either side. The outer membrane (OM) has a different chemical construction than the inner, plasma membrane (PM). Both contain protein molecules (P) which can have enzymatic and/or structural functions and phospholipids, which have polar head groups (depicted as circles in B) and two fatty acid chains. In the OM these are mostly limited to the inner leaflet. The outer leaflet is mainly composed of lipopolysaccharide (LPS) which has a higher number of fatty acid chains, a more elaborate "polar head group", and external, branched, carbohydrate chains. These can have a high negative charge density and bind metal ions, an ability demonstrated by the more intense staining of the outer leaflet on the cell in A. Bar = 100 nm.*

The outer membrane also has various types of proteins embedded in it which are usually present as high copy numbers of only a few different types [15,4]. These proteins may either span the membrane or be embedded in either face. Generally there are three types of proteins present: (a) the "porins" which form water-filled channels through the outer membrane to allow the passage of hydrophilic solutes, (b) proteins which allow the binding of specific molecules for transport into the cell, and (c) those which serve to bind the outer membrane to the peptidoglycan layer. The latter, known as lipoproteins, are very small polypeptides (made up of approximately 58 amino acids) with a hydrophobic end that anchors the molecule to the outer membrane and a hydrophilic end bound to the peptidoglycan layer [15]. In *E. coli* only one third of the lipoprotein molecules are actually covalently bound to the peptidoglycan, the rest are associated with the bonded variety in a 2:1 stoichiometry, forming trimers [13].

The peptidoglycan layer of gram-negative bacteria is much thinner than that of their gram-positive counterparts (only 3 nm for *E. coli* as opposed to about 25 nm for *B. subtilis*). That of *E. coli* K-12 is also less highly cross-linked than that of *B. subtilis* 168 (25% vs. 50%) but is chemically similar with respect to reactive sites for metal binding [16].

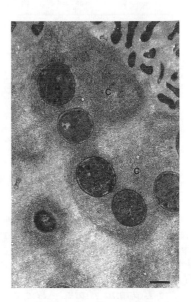

Figure 4.   *TEM Image of Cross-sectioned Cyanobacterial Cells from Mono Lake, California. The cells (arrows) are enveloped in a thick layer of fibrous capsular material (C). Bar = 1 μm.*

2.3   Other cell surface structures

Bacteria may possess other surface structures such as pili or fimbriae, flagella, spinae, S-layers and capsules. Of all these structures, the importance of capsules to metal binding has been most thoroughly studied. Capsules (Figure 4) are highly hydrated, loosely arranged structures which extend outwards from the cell wall matrix. They are typically anionic and have great potential for scavenging metals from the environment [17]. In contrast, S-layers are highly organized surface layers made up of monomers of a single (usually) protein species that are arranged in a regular pattern, most commonly hexagonal or tetragonal [18]. In cross-section (Figure 5), they appear as a layer with a regular periodicity which completely encloses the cell. Few studies on their metal sorption capacity have been performed. Since a description of the chemical nature of these structures is intimately related to a discussion of their metal binding characteristics, they will be dealt with in detail later in this review.

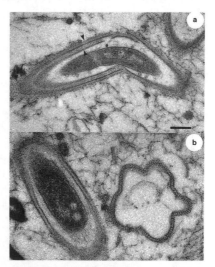

Figure 5.   *S-layers are Often Found on Cells in Environmental Samples. They appear, in cross-section, as patterned layers external to the bacterial cell wall and can often be elaborate; cells may posses more than one S-layer. These TEM images show bacterial cells from samples collected on the Bahama Bank. The cell in A possesses two S-layers (arrows). Bar = 500 nm. Two different S-layer structures are contrasted in B. The remnant wall of a clover leaf-shaped cell shows a fence-like pattern while that of the adjacent cell has a more typical periodic pattern made of darkly-staining spots on the array. Bar = 200 nm.*

2.4 Archaeobacterial walls

An appreciation of the variety and complexity of wall structures found within this group of microorganisms has only begun to develop. Consequently, little is known of their interaction with metal ions. Four types of archaeobacterial wall polymeric structures are known [19]: (1) pseudomurein, a heteropolymer similar to peptidoglycan but with N-acetyltalosaminuronic acid instead of N-acetylmuramic acid (e.g., *Methanobacterium*). (2) Walls containing methanochondroitin, a molecule similar to that found in the connective tissue of higher animals (e.g., *Methanosarcina*). (3) Paracrystalline proteinaceous surface arrays (S-layers) as the only component overlying the cell membrane (e.g., *Halobacterium*). The other wall types already mentioned may have an S-layer in addition to the components described. (4) The genus *Thermoplasma* is made up of members that have no external wall layers at all; they are bounded simply by their plasma membrane.

Some archaeobacteria grow as filaments surrounded by a resilient sheath (Figure 6) which seems to have an affinity for certain metal ions (e.g., *Methanosaeta, Methanospirillum*). For example, the sheath of *Methanosaeta* (*Methanothrix*) *concilii* seems to preferentially bind Zn^{2+}, Co^{2+}, Ni^{2+} and Fe^{2+} ions [20]. This is not surprising when one realizes that these ions figure prominently as cofactors in the cells' metabolism.

3. BIOFILMS

The natural environment is dynamic and microorganisms must continously readapt themselves to chemical and physical changes. For bacteria, the different states of matter (gas, liquid, or solid) are very important since their junctures produce unique hybrid microenvironments. These interfaces between solid and gas, liquid and gas, and solid and liquid possess qualities that are subtly different from either of the two bulk phases which flank them. Of particular interest to this discussion is the solid/liquid interface.

The unique physical conditions present at the solid/liquid interface, most notably the "hydrodynamic boundary layer" (a layer of still water immediately adjacent to the solid surface which varies in thickness with changes in the velocity of the water flowing past [21]), produce conditions conducive to the growth of microorganisms. Indeed, the majority of microbial activity in aquatic environments appears to be associated with submerged solid surfaces [22]. Over 90% of the total bacteria in natural aquatic environments can be found attached to solid surfaces and in the monolayer at the water/air interface [23]. These biofilms consist of a complex community of interdependent microorganisms, dominated by bacteria (Figure 7). Many studies have been undertaken in order to establish how the formation of a biofilm is initiated and how it alters as it is subjected to environmental changes. Due to the vast quantity of information available, only a brief overview will be given here. The interested reader is referred to Marshall [24], Fletcher [25], and Costerton [26] for detailed reviews of the subject.

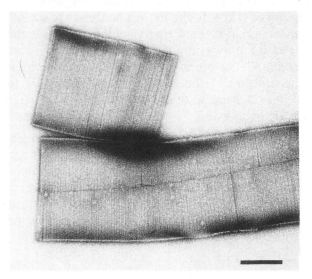

Figure 6.   *The Sheath of <u>Methanospirillum hungatei</u> is made up of a Series of Hoops Which are Stacked Together to Form a Tube-like Structure within which the Filamentous Chains of Cells Reside. This TEM image of negatively stained, collapsed, sheath shows the striated appearance arising from its structural organization. The proteinaceous sheath fabric is highly resistant to chemical degradation and can bind metal ions. Bar = 500 nm. (Micrograph provided by Dr. G. Southam, Dept. of Microbiology, University of Guelph).*

Most solid materials, such as rocks, have a net electrical surface charge and, at the same time, can exhibit a certain degree of hydrophobicity. This patchy charge distribution is due to lattice defects and fractures (which can become highly charged), while other regions of the surface are electrically neutral. The size and number of each region gives the exposed surface its overall hydrophilic/hydrophobic character.

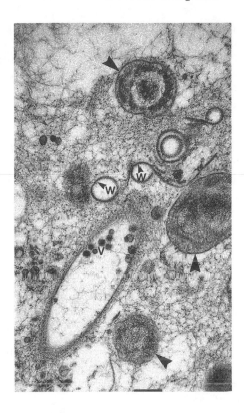

Figure 7.   **When Natural Biofilms are Thin-sectioned and Examined by TEM, They are Revealed as Complex Structures Composed not only of Living Cells, but also Cell-derived Polymers and Cell Remains.** *This image of a marine biofilm from the Pacific coast of British Columbia shows several bacterial cells (arrows) accompanied by the remnants of cell walls (W). The S-layer is all that remains of a bacterium that may have been attacked by a virus; numerous viral particles (V) appear to be present within this empty shell. The entire biofilm is permeated by polymeric, mostly capsular material (C) produced by the cells within it. Bar = 200 nm.*

As soon as any solid surface contacts an aqueous phase it quickly acquires a layer of ions and organic matter. This is inevitable since absolutely pure water can not exist in a natural environment. Neihof and Loeb [27] have shown that a "conditioning film" of ions and macromolecules will form on glass, clay and calcite particles almost immediately after their immersion in sea water. Subsequent work [28] has provided confirmation for the existence of conditioning films. This active concentration of inorganic and organic compounds on solid surfaces, above and beyond the concentrations found in the overlying water, provides a highly satisfactory nutrient source for the growth of bacteria. Once they contact the nutrient-enriched surface, bacteria adhere to and colonize it, leaving behind the relatively less appealing bulk water of their surroundings. All surfaces in natural systems are modified by microbial attachment.

Although the idea of nutrients accumulating at solid/liquid interfaces is attractive, this concept is an oversimplification of the complex interactive conditions which lead to microbial colonization of a submerged solid surface [29]. Other factors have been implicated in the encouragement of bacterial adherence to surfaces. In highly turbulent fluids such as mountain streams and high-flow water pipes, biofilm formation could be an attempt by bacteria to shelter themselves in regions of relatively low shear [30,31]. Even in high-velocity fluids the drag along solid surfaces reduces shear. In fact, the eventual development of biofilms should induce even more drag, less shear, and a more unchanging microenvironment in which the cells can grow [23].

In a mechanistic sense, the actual adsorption of bacteria to solid surfaces and eventual biofilm development is quite complex. Once the cells encounter a solid surface, weak bonding events are believed to occur which allow the cells to be held in place until they form a more permanent means of adhesion. It is a two stage event consisting of reversible and irreversible adhesion [32]. The former results from electrostatic repulsive forces between the two negatively charged surfaces (each surrounded by a layer of cations) being balanced by van der Waals attractive forces which result from the innate properties of the two surfaces. This leads to a net attraction between the bacterium and the surface over discrete distances which vary according to the osmotic strength and ionic composition of the aqueous phase [30]. As a result the cell is held near the surface at a distance of 7 to 10 nm. Since, at this stage, it is a loose attraction, the cell still exhibits Brownian motion and can move laterally [33]. Somehow this close proximity triggers the bacterium to respond by producing extracellular polymers or in some other way changing its surface to allow it to adhere firmly ("irreversibly") to the surface. The production of extracellular polysaccharide seems to be the prevalent reaction. By this means cells can glue themselves so firmly that they can only be removed by excessively strong shear forces [34]. The end result is that a biofilm consists of a large number of (mostly) bacterial cells of differing species embedded in a highly hydrated polymeric matrix.

Since biofilms consist mainly of polysaccharide polymers it is the acidic chemical groups of this extracellular matrix which will have a major bearing on the metal binding capacity of the film. Biofilms are ubiquitous and formidable entities on surfaces in the natural environment and therefore must have a major impact on the chemistry and microbiology of the water immediately overlying them. It is therefore essential to learn more about how they influence these environments and, more importantly, the extent of the ability they have to immobilize soluble pollutants such as metals. Although most biofilm studies have been done on "open" water systems, the subsurface contains an ample variety of bacteria at concentrations ranging from 10^3 to 10^6 cells per square centimeter [35,36]. These subsurface biofilms must have a substantial impact on the transport of metals in groundwater.

4.   METAL ION BINDING BY BACTERIAL CELLS

In nature bacteria are surrounded by aqueous solutions containing many different organic and metal ions. This is true even for those bacteria that live in seemingly dry conditions under the mineral varnishes of soil aggregates and rocks in both hot and cold desert regions. For at least part of the year (during which the cells are active) they are surrounded by a thin film of water. Bacteria require water not only to maintain cellular integrity (through turgor pressure) and proper metabolic functioning, but also to carry nutrients to the cell. They rely entirely on diffusion in their immediate local environment to obtain necessary compounds and to disperse metabolic waste products [37]. It is the surface layers which have first exposure to the diffusible components of the environment surrounding them. Knowing this it would be unreasonable to assume that bacteria would not have metal ions and other environmentally-derived compounds intimately associated with their surfaces. The bacterial cell surface provides a large surface area for interaction with the metal ions in which the cell is constantly being bathed. In fact, of all cellular life forms, bacteria have the largest surface area to volume ratio for these sorts of interactions [37]. Although in most natural environments metal ions are present in low concentrations, bacterial cells show a remarkable ability to concentrate metal ions out of aqueous solutions.

4.1   The influence of cell surface physicochemistry

Regardless of whether bacterial walls are of the gram-positive or gram-negative variety they tend to have a net negative charge at circumneutral pH. Those of *E. coli* have a remarkable capacity to accumulate metal which is attributable to this charge character [38]. Their negative charge is contributed mainly by phosphoryl groups present in the core oligosaccharide and the N-acetylglucosaminyl residues in the lipid A moiety of the lipopolysaccharide molecule [14,39]. Although there are carboxyl groups present in the 2-keto-3-deoxyoctonate residues of the lipopolysaccharide molecule, studies with *E. coli* K12 (a mutant lacking the O-antigen portion of its lipopolysaccharide) have shown that only one of three carboxyl residues is available for metal binding, the others being cross-linked to amino groups within the molecule [39]. The peptidoglycan layer of gram-negative bacteria is also capable of binding metals. Its capacity on a percent dry weight basis is similar to that of gram-positive bacteria but on a per cell basis it is less since there is less of this polymer present [16].

Gram-positive surfaces have an even higher capacity to bind metal ions [7] than those of gram-negative bacteria. The prime sites of cation binding in the wall of gram-positive bacteria seem to be the carboxyl residues of peptidoglycan and teichuronic acids, and the phosphoryl groups of teichoic acids. The contribution of each to the overall metal binding capacity of the cell depends on how much of each polymer is present. For example, *B. subtilis* cells grown in media with sufficient magnesium and phosphate have walls consisting of 54% teichoic acid and 45% peptidoglycan [9]. Extraction of the teichoic acid shows that the majority of the metal associated with the cell was bound to the peptidoglycan portion of the wall [9]. On the other hand, *B. licheniformis* cells which have walls consisting of 26% teichuronic acid, 52% teichoic acid and only 22% peptidoglycan lose most of the wall-associated metal when the two acid polymers are removed. This indicates that when gram positive walls have low peptidoglycan percentages, the secondary polymers control the wall's charge density and, therefore, its metal binding capacity [11]. The wall physichochemistry of gram-positive bacteria allows the cells to concentrate metals out of dilute solution. *Micrococcus luteus* was able to scavenge Sr^{2+} from a 500 ppm solution until 25 mg Sr was bound per gram of cells [40], representing a 50-fold increase in concentration.

So far the mechanistic details of metal binding to bacterial walls are known only for *E. coli* K12 and *B. subtilis* 168. However, since the heavy metal stains routinely used for electron microscopy bind to the surface of most bacteria, representing a massive diversity of ultrastructural and chemical types, the ability to bind metals is not an isolated peculiarity confined to these two organisms. It is a widespread reflection of the general anionic nature of bacterial cell walls.

Beveridge and Murray [12] proposed at least a two-step mechanism for the deposition of metal ions in the bacterial cell wall. The first step involves a stoichiometric interaction between metal cations and active sites within the wall. This interaction provides a nucleation site for the deposition of more metal from solution. The metal aggregate grows within the wall until it is physically constrained by the size of the intermolecular spaces in the wall fabric itself. As a result, metals deposited in the wall are not easily remobilized by water or replaced by protons or other metal ions.

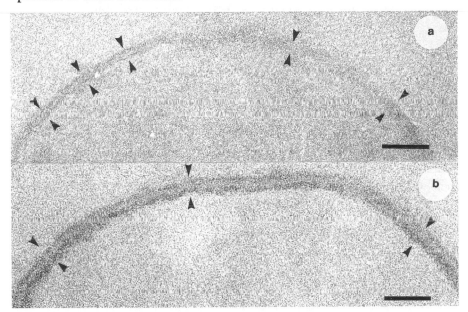

Figure 8.   *Transmission Electron Micrographs of Thin-sectioned Metabolizing (A) and Sodium Azide-uncoupled (B) Bacillus subtilis Cells. The cells in B are no longer releasing protons (a normal consequence of metabolism in living bacteria) which compete with metal ions for binding sites within the wall. Contrast was provided exclusively by the UO_2^{2+} ions bound by the cell walls (delineated by arrows). Bars = 100 nm.*

In natural environments, cells are faced with a wide range of ions which may compete with metals for charged sites within the wall fabric. However, the greatest competition for binding sites within the walls may come from the cells themselves. Experiments conducted with *B. subtilis* have shown that protons that are pumped into the wall fabric by the membrane-induced proton-motive force during metabolism can effectively compete with metal ions for anionic sites within the walls.

As a result, less metal is bound by living than by non-living cells or those in which the plasma membrane has been uncoupled [41]. This implies that, in the natural environment, dead cells and their various extruded polymers could play a more important role in metal immobilization by passive processes than metabolizing cells. As metals denature and inactivate cellular autolysins [42], the wall material will persist for long periods of time, continuing to provide sites for nucleation of minerals to occur.

4.2   Metal ion interaction with bacterial capsules

Most capsules are composed of polysaccharides arranged in simple or branched chains made up of repeating units of 2 to 6 different sugars. The predominant chemical species present are uronic acids which may make up to 25% of the capsule polymer [17] yet the bulk of the capsule is made up of water (about 99% by weight). The result is a highly hydrated gel-like structure which can interact intimately with solutes present in the surrounding water. As has already been indicated, capsular material is a major component of biofilms and, as a consequence of the ubiquity of these microbial communities, capsular material is also a major component of the total organic fraction in natural systems.

The capsule acts as a sort of "buffer zone" between the cell and its environment and is the first structure encountered by metal ions when they are in the vicinity of the cell. Due to their chemical nature, which resembles that of a cation exchange resin, capsules are ideal cation scavengers and can accumulate large quantities of metal ions [43]. *Zoogloea ramigera*, a bacterium that forms an extensive capsule, was found to contain 25% by weight of metals after growing in sewage sludge [17]. Other studies indicate that up to 3 mmol Cu/g dry weight of polymer can be accumulated by these cells [44].

The major functional groups of capsules responsible for metal binding are carboxyl and hydroxyl groups. Of the two chemical groups, carboxyl groups are the most active (e.g., the gamma-glutamyl groups of the *B. licheniformis* capsule [45]). For these charged polysaccharides, metals are usually bound by cross-bridging between anionic groups. This neutralization of charge by the metal ions often results in coprecipitation of the metal-polymer composite resulting in floc formation [17,46]. Thus, metal binding to the extracellular polymers of planktonic bacteria can provide an effective means of transporting metals from the aqueous phase to the sediment in natural environments.

Although most capsular polymers are charged, uncharged polymers can also be found. In this case, weak electrostatic interactions between metals and hydroxyl groups become important. The affinity of uncharged polysaccharides for the metal ions tends to decrease with increasing radius of the hydrated metal ion. In general, an acid-base reaction is involved which results in the liberation of protons [47,43]. Such reactions appear to be significant in the corrosion of metal surfaces by bacteria.

Due to their exceptional ability to accumulate metals, capsules have been advocated as an additional design strategy by which bacteria protect themselves against toxic metal concentrations. There is a fine line between the concentrations at which metals are essential or toxic. Since metal concentrations often fluctuate widely in the natural environment, it is necessary for bacteria to have a mechanism by which the concentration of metal that actually reaches the cell surface can be controlled [17]. Ditton and Freihofer [48] have shown that unencapsulated mutants of *Klebsiella aerogenes* were killed by metal concentrations to which their encapsulated counterparts were resistant.

The presence of metals can even enhance capsule formation. An increase in the concentration of Cr^{3+} led to increased polymer production by a coryneform bacterium isolated from Cr polluted water [49]. Capsule composition can also be influenced by the metals present. The removal of Ca^{2+} from the growth medium of *Azotobacter vinlandii* resulted in an increased proportion of mannuronic acid in the exopolymers produced by this bacterium [50,51]. It is apparent that bacteria can change the chemical nature and amount of material of their capsules in response to changes in external metal type and concentration. For example, the degree of acetylation of the polymers strongly influences the selectivity of metal binding, as has been recently demonstrated by Geddie and Sutherland [52] with a variety of bacterial polysaccharides. This gives them greater flexibility in dealing with the types and concentrations of metals they are faced with.

4.3   The metal-binding ability of S-layers

Few, if any, quantitative analyses of metal binding ability of an S-layer-carrying organism have been undertaken although qualitative assessments of the ability of these structures to bind cations have been made. Some S-layers require the presence of certain cations for stability such as Ca^{2+} or Sr^{2+} [53,54,55] or Mg^{2+} [56]. However, the ability of an S-layer to act as an ion-sequestering structure without having a structural role for these ions has not been widely documented. Under reducing conditions, the proteinaceous sheaths of *Methanospirillum hungatei* and *Methanothrix concilii* avidly bind nickel [57], a metal that is required for the activity of several key enzymes [58]. These structures can concentrate nickel from solution and may act as a reserve for this metal [57].

5.   MINERAL FORMATION ON BACTERIAL CELLS

The matrix of the bacterial wall provides a special environment for the nucleation and growth of metal aggregates. This results in an effective immobilization of metal ions and may be an important factor in the transport of metals. For example, in lake sediment [46], bacteria concentrate metals, which may undergo subsequent geochemical processing, transforming them into particulate minerals (e.g., metal oxides and hydroxides, carbonates, sulfates, etc.) as either amorphous or crystalline precipitates.

The idea that bacteria could provide nucleation sites and act as templates for the formation of minerals provides some explanation for the existence of microfossils [59]. Beveridge and coworkers [60] undertook low temperature diagenesis simulations of geological processes to see what types of minerals would form on metal-loaded *B. subtilis* walls exposed to artificial sediments containing either calcite, quartz or a 1:1 mixture of the two in the presence of either sulfur or magnetite as redox buffers. They found that when sulfur was present sulfitic minerals formed. However, in the absence of sulfur the dominant mineral types were phosphates with the phosphate being provided by the bacterial walls. The mineral composition seemed to depend on the type of metal sorbed to the walls and the mobile ions present in the simulated sediment.

The actual formation of highly mineralized bacteria resembling intact microfossils as seen in natural environments [61] was successfully simulated by Ferris and coworkers [42]. Whole cells were artificially aged for up to 150 days at 70°C in the presence of silica. However only those cells that had been pretreated with iron remained structurally intact and recognizable as bacteria. It was concluded that the binding of heavy metal ions such as iron by bacterial cells inactivated cell wall hydrolyzing enzymes and was an important factor in preserving cellular shape, contributing to the fossilization of microorganisms.

Laboratory experiments that simulated the dilute conditions of soil solutions have shown that metallic silicates of low crystallinity can be nucleated on the surface of *B. subtilis* cells, either at pH 5.5, 4.5, or 8.0.

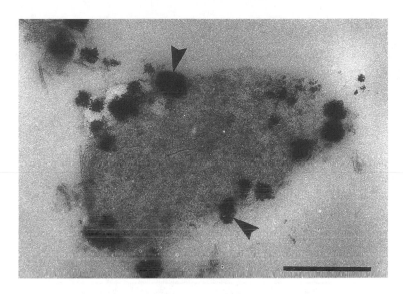

Figure 9.   *To Study the Formation of Mineral Aggregates on Bacterial Cell Walls and Their Subsequent Diagenesis, Cells were Loaded with Uranium and Aged for Ten Days in the Presence of Elemental Sulfur as a Redox Buffer. Under these conditions, extracellular uranium phosphate microcrystals (arrows) were formed. The phosphorus came from the cell wall of the Bacillus subtilis cells used for the study. Bar = 500 nm.*

This process was particularly evident at pH 4.5 when heavy metals (Cd, Zn, Cu, Cr, and Ni) were combined with Al and Fe (which prevented autolytic degradation of the cell walls). In all circumstances, the bacterially-mediated mineral phases were more abundant, diverse, and smaller than in abiotic controls. In addition, the metal ions were more efficiently incorporated into the mineral structures than under inorganic conditions [62]. The mechanism of silicate binding to a surface that is predominantly electronegative was elucidated by chemically modifying the carboxylate groups in the wall fabric [63]. Silicate anions bound to the bacterial surfaces by electrostatic interaction with native $-NH_3^+$ residues in the wall or with previously wall-bound multivalent cations that had some charge still available. This last mechanism (cation-bridging) involves the formation of a ternary wall-cation-silicate complex and is presumably the predominant means of mineral nucleation by negatively charged bacterial surfaces in the natural environment. The significance of these results can be appreciated when considered in the context of sediment environments, where the microbial input is subtle and is entwined with a complex variety of other organic and inorganic components, chemical reactions, and physical phenomena. It is important to recognize that microorganisms might mediate the neoformation of secondary silicates in soils and sediments; a role that has not been previously considered.

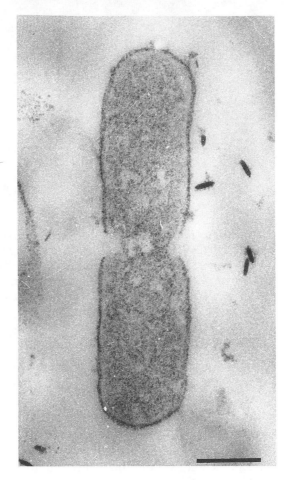

Figure 10.   *Low-temperature Laboratory Diagenesis Experiments Yielded Structures Resembling Bacterial Microfossils. An essential component of the process was the binding of iron by the autolytic enzymes present in the cell wall. This inactivated them, preventing degradation of the peptidoglycan network. Binding of metal ions by bacterial cell surfaces can therefore ensure their persistence in the environment for a long period of time and provides a means of immobilization for these ions. Bar = 500 nm.*

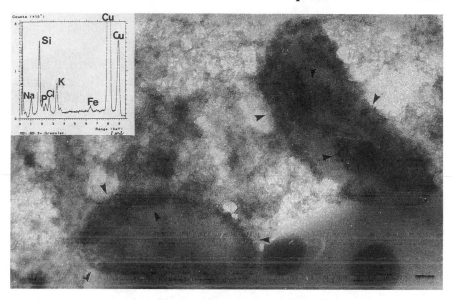

Figure 11.   *Transmission Electron Micrograph of a Whole Mount
                  Preparation of* <u>Bacillus</u> <u>subtilis</u> *Cells Mineralized by
                  Metal-silicate Crystallites. Arrows indicate points of
                  contact between the bacteria and the silicate phases they
                  have caused to form by providing nucleation sites. The
                  inset shows a typical energy-dispersive x-ray spectrum
                  of the mineralized bacterial surface, where Cu peaks
                  correspond to the Cu grid used for TEM, P peak to the
                  cell wall components, and K peaks to the cytoplasmic
                  contents. Si, Fe, and Na were cell-bound from the
                  reaction mixtures. Bar = 100 nm.*

5.1   *In situ* observations

        Examination of samples from many different sites, in a wide range
of natural environments, invariably reveals evidence of microbial metal
binding and mineralization. The easiest and quickest way to verify this
is to examine unstained samples by electron microscopy. In such
samples aggregates of usually amorphous minerals are often arranged to
conform to the outline of a recognizable microorganism. The identity of
the minerals can be established by analytical methods such as energy-
dispersive x-ray spectroscopy (EDS) and crystallographic techniques
such as selected area electron diffraction (SAED). The morphology of
the microorganism and the arrangement of its enveloping layers can be
examined by subsequent staining of the sample with heavy metals such
as uranium and lead, which lend contrast to biological structures. This

provides confirmation that mineralization is taking place on or within the microbial cells present, the majority of which are bacteria.

Some bacteria, such as *Leptothrix* and *Sphaerotilus*, both ensheathed in amorphous tubes composed of loosely arranged polymers, are able to build up immense concentrations of manganese and iron oxides, respectively [57]. In fact, for *Leptothrix discophora,* the sheath possesses enzymatic oxidizing activity [64,65]. These can become significant agents for metal immobilization and transport in the environments these bacteria inhabit. Extensive manganese oxide deposition has been found to occur in microbial mats in hot springs in Yellowstone Park [66] as well as a wide range of other environments including lakes [67] and the ocean floor [68]. Also, nucleation of iron-silica crystallites on bacterial surfaces occurs in the sediment of an acidic hot spring in Yellowstone National Park [61] and in water column samples from the Rio Negro in Brazil [69].

In situ examination of microbial biofilms confirms their importance in metal binding and mineralization in the natural environment. Ferris and coworkers [70] studied metal-contaminated lake water in which biofilm metal adsorption (per cm^2) was up to 3 orders of magnitude greater than was present per mL of the overlying water. Throughout the 17 week study period, there were 10 to 100 times more bacterial cells present in the biofilms (per cm^2 of surface) than per mL of the bulk fluid phase. It is likely that the extracellular polymers which knitted the cells together in the biofilm also contributed to the greater metal sorption. Indeed, electron microscopy and EDS revealed abundant iron-rich precipitates, tentatively identified as ferrihydrite, throughout its polymeric framework. Other forms of Fe, like hematite and goethite, were also found encrusting microbial cells in lake sediments under the influence of acid mine drainage [71].

Perhaps the most striking and widespread examples of the ability of biofilms to effect mineralization are those seen in microbialitic structures such as stromatolites and thrombolites. These are laminated or open porous carbonaceous structures, respectively, which have been formed by the growth and subsequent mineralization of oxygenic photosynthetic phototrophs; predominantly cyanobacteria. This special case of microbial mineralization is explored further in the following section.

5.2   Fayetteville Green Lake

Fayetteville Green Lake is a small, meromictic lake situated within Green Lakes State Park near Fayetteville, New York. It is fairly deep (approximately 55 m at the deepest point) and has an unusual chemistry that has attracted researchers for well over a century. A thorough review of the fascinating history of research of this lake along with an investigation of its microbially-driven geochemistry is provided by Thompson and coworkers [72]. Reviews of the lake chemistry and physical characteristics are given by Brunskill and Ludlam [73] and Torgersen *et al.* [74].

The unique chemistry of this lake, particularly the high carbonate (3.57 - 7.43 mmol/L [74]), sulfate (11.66 - 15.09 mmol/L [73]) and very low ammonium and iron concentrations [73] ensure that eukaryotic phototrophs are at a disadvantage compared to prokaryotic phototrophs such as cyanobacteria and anaerobic purple and green sulfur bacteria. Prokaryotes are extremely efficient scavengers of nutrients in oligotrophic environments [75] and can successfully out-compete eukaryotic phototrophs in environments like Green Lake, thereby becoming the dominant phytoplankton species. In the anoxic zone, which begins at the permanent chemocline in this lake, green and purple sulfur bacteria thrive on the high levels of sulfide emanating from the lower regions of the lake.

The lake also contains extensive carbonate bioherms (modern stromatolites/thrombolites) of which the origin was long a mystery until the discovery of a small (<1 μm cell diameter) unicellular cyanobacterium. This appears to be the only significant phytoplankton species in the lake and is invariably found in association with bioherm material when viewed by light and electron microscopy. In 1963 W.H. Bradley noted small bacterial cells occluded within some calcite grains from the bottom sediment of Fayetteville Green Lake. Subsequently, Thompson and coworkers [72] demonstrated that the occluded cells were in fact small cyanobacterial cells belonging to the *Synechococcus* group.

Synechococcus thrives in the oxygenated zone of the lake. Its growth is seasonal and peaks during the warmer months of July and August. Interestingly, it is during these months that calcite mineralization is greatest, yet the exact reason for calcite precipitation was at first unclear. At the circumneutral pH of the lake, precipitation should be in the gypsum solid field; calcite precipitation requires a more alkaline pH. *Synechococcus* initiates carbonate deposition within its immediate vicinity by the photosynthetically-driven alkalization of the microenvironment surrounding the cell. In this process, bicarbonate is taken into the cell where it is oxidized by carbonic anhydrase to form CO_2 and OH^- [76]. The CO_2 is incorporated into organic molecules via the Calvin-Benson cycle and ultimately is used to provide molecular building blocks and energy (in the form of stored carbon compounds such as glycogen). The OH⁻ is released by the cell and produces an alkaline microenvironment immediately peripheral to the cell surface [77,76]. This alkalization moves calcium precipitation from gypsum into the calcite field and provides for the marl sediments and calcareous bioherms [78].

Laboratory simulations using lake water and isolated *Synechococcus* cells confirmed alkalization of the water by these cells and revealed their ability to promote epicellular calcite mineralization [78]. In this study the mechanism for calcium sequestering on the cell surface was also addressed; the surface of *Synechococcus* cells appeared to have an affinity for calcium ions.

Recent structural studies of *Synechococcus* reveal that it has a gram-negative wall ultrastructure but its outermost surface layer is an S-layer with a hexagonal symmetry. Subsequent laboratory studies have revealed that the S-layer acts as a template, providing regularly-spaced, chemically identical sites for the nucleation of minerals [79]. Calcium ions bind in the diamond-shaped holes of the array and are joined by sulfate ions to form gypsum crystals. In the early stages of mineral formation the S-layer pattern is still clearly visible. As the pH continues to increase, the preferred mineral form becomes calcite, which also grows on the S-layer, eventually obscuring its pattern. As the surface layers of the cells become encrusted with minerals, they are shed by the cells and replaced by new surface layers which become mineralized in turn [79]. This shedding activity is essential to protect the cell's ability to grow and divide since both require an increase in cell size. Division occurs once every two days while the cell surface can become extensively mineralized within eight hours [79]. Eventually, when the cells reach a stationary growth phase, their metabolic activities slow down and shedding of surface layers ceases. The cells become entombed in calcite and, if growing as a biofilm on the lake shore, contribute to bioherm growth. Cells that are suspended in the open water column gradually drift down to contribute to the extensive marl sediment at the lake bottom. This is manifested by a whitening of the lake water that occurs in the warm part of every season.

Fayetteville Green Lake provides an excellent example of the ability of microorganisms to effect overt physical changes in their environment as a result of their involvement in mineralization processes. The appearance and growth of the carbonate bioherms and the substantial thickness of marl sediment at the lake bottom provide evidence that microbial life has played (and is still playing) a major part in the formation of geologically significant structures [72,78]. The discovery of the mechanism of bioherm formation in the lake has far-reaching implications for the origin of similar structures found in the fossil record and in other parts of the modern world.

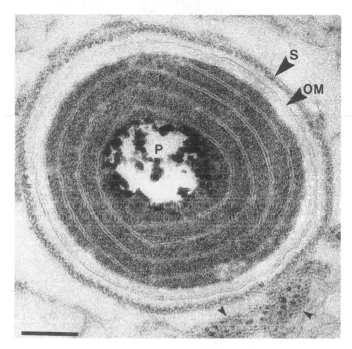

Figure 12.   *Cross-section of a <u>Synechococcus</u> Cell Isolated from Fayetteville Green Lake. The cell is completely surrounded by an S-layer (S), which it sheds, together with outer membrane material, into the surrounding water (arrows). Under the S-layer is the outer membrane (OM). The concentrically-arranged membranes within the cell are the thylakoids, within which the photosynthetic and respiratory electron transport chains are located. Most of the cell center is occupied by the remains of a large polyphosphate granule (P), used to store phosphorus. Bar = 200 nm.*

6. METAL ION TRANSPORT AND THE IMMOBILIZATION OF TOXIC HEAVY METALS

Metals are an integral part of the Earth's crust and are present (through complexation with phosphate, carbonate, silicate, hydroxide and sulfide ions) mostly as insoluble precipitates and minerals. Natural geochemical dissolution, weathering and microbial leaching [80] are

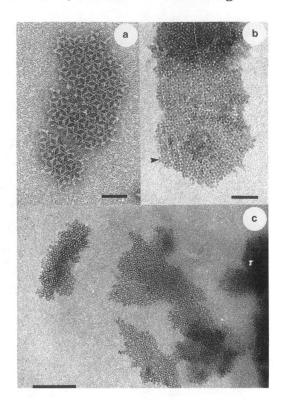

Figure 13.   *When Whole Mounts of S-layer Fragments from Synechococcus Cells are Negatively Stained, the Dark Stain Pools in the Holes of the Array, Outline the Shape of the Protein Network (A). The overall hexagonal symmetry is apparent, in which the large diamond-shaped holes of the array spiral around a central pore or pit to form a pinwheel pattern. Bar = 50 nm. Mineral formation (B) in lake water begins in the large holes of the Synechococcus S-layer as Ca ions bind to negative sites within them and attract sulfate ions to form gypsum ($CaSO_4 \cdot 2H_2O$) or carbonate ions to form calcite ($CaCO_3$), depending on the pH of the surrounding microenvironment, which is controlled by the cells. The pinwheel pattern of the S-layer can still be seen (arrows). Bar = 100 nm. Even S-layer that has been shed from the cell (C) can act as a template for mineral formation. Shedding is a natural response of the cells to avoid total encrustation by minerals, which would interfere with growth and division by impeding ion transport into and out of the cells and obscuring active sites of enzymatic activity. Bar = 200 nm.*

responsible for their release in a soluble form and account for their natural concentrations in freshwater systems. Perhaps a greater load of metal ions enters the environment as a result of human activity such as mining, heavy industry and metal refining. Heavy metal pollution has become a major concern in recent years and it is important for us to understand what happens to the metals in the natural environment. This will hopefully lead us to ways of containing and possibly recycling these metals for future use. It is here we expect to find that bacteria may play an important role in impeding the transport of metals in natural systems. The fact that bacteria can bind metals is perhaps not surprising but the tremendous quantity they can accumulate, especially from dilute solution, is remarkable. The major factors responsible for this ability are their small size and the physicochemical attributes of their surfaces. Their large surface area-to-volume ratio promotes the efficiency of the immobilization process. The fact that bacteria are ubiquitous and present in enormous numbers allows them to have a significant impact on metal transport and immobilization processes in natural systems.

In the water column of temperate climate lakes, we envision planktonic microorganisms and their remains as forming a season-variable light rain of microscopic particles which cleanse the waters of dilute, soluble, toxic heavy metals, forming aggregates which, over time, increase in size through flocculation. These settle to the sediments and account for the high concentrations of heavy metals found in this fraction.

Biofilms, on the other hand, form heterogeneous but stable outer layers on hard surfaces such as rocks that can approach millimeters in thickness. They avidly concentrate toxic metals and become, themselves, mineralized over time. Certainly, as in the Fayetteville Green Lake system, cyanobacterial communities and their alkalization can account for tremendous mineralization and can effectively control the type of mineral that is formed.

Although we have been stressing the importance of bacteria as the major presence responsible for metal binding in natural systems, it is prudent to point out that there are other components present in the environment that are capable of presenting a significant impediment to the mobility of metal ions. Other biological particulates are essentially made out of the same "stuff" as bacteria and should also interact with toxic metals. Some of the most active components are soluble organics such as humic and fulvic acids. Clay-sized minerals (<2 μm) also play a prominent role in metal immobilization, particularly electronegatively charged phyllosilicates and Fe and Mn oxyhydroxides.

Walker and coworkers [81] were interested in finding out how bacteria and clay-minerals interacted with each other and how this interaction could affect the metal binding capacity of each component. Does this interaction enhance or impede the metal binding capacity of the system? This interest arose from the recognition that the laboratory simulations with bacterial surfaces were an oversimplification of the conditions actually present in the natural environment. Once the capacity of bacteria to bind metals and the mechanisms by which the binding occurs was fairly well understood [7,57], and the associations between clays and metals were well recognized [82,83], it became important to understand how these two components interact to immobilize metals.

Walker and coworkers [81] used *B. subtilis* walls and *E. coli* envelopes as representative microbial particulates. These were mixed with each of two types of clay-minerals, kaolinite and smectite (montmorillonite). The metal binding ability of each separate component as well as that of composites was determined. Bacteria-clay composites formed best when mixed in a 1:1 proportion (on a dry weight basis) and the clays showed a preference for edge-on orientation with the organic fraction. This indicated that binding occurred between the negatively charged walls and envelopes and the positive edges of the silicates. The addition of multivalent metal ions increased the incidence of binding since planar surface orientations were also involved. This indicated that the metal ions were acting as bridges between the two particulate components.

The comparative metal binding abilities of the components in terms of metal bound per dry weight of clay, walls, envelopes, or composites was determined [81]. The results indicated that gram-positive walls bind a greater quantity of metal than gram-negative envelopes and that both bind more metal than the layered silicates. The sum of the metal binding capacity of each of the individual components exceeded that of the bacteria-clay composites. This was due to the masking of a proportion of the available metal binding sites during composite formation. On a dry weight basis the envelope-clay and wall-clay mixtures bound 20% to 90% less metal than equal amounts of the individual components. The organic portion of the composite accounted for most of the binding capacity [81].

These experiments were extended by Flemming and coworkers [84] who wished to determine what conditions were necessary to remobilize the metals bound to the bacteria-clay composites. This could give some indication of the nature and strength of the forces which bound the metals to the walls, envelopes, and clay minerals. Several concentrations of nitric acid, calcium nitrate, ethylenediaminetetraacetic acid (EDTA), soil fulvic acid and lysozyme (an enzyme that degrades the peptidoglycan in bacterial cell walls) were tested. The remobilization of the sorbed metals depended on the physical properties of the organic and clay surfaces and on the character and concentration of the leaching agents. Although each leaching agent was effective in remobilization of certain metals (in this study Ag^+, Cu^{2+}, and Cr^{3+} were used), the greatest mobility was achieved at acidic pH or with elevated Ca^{2+} levels (160 ppm). Under all conditions, less metal was resolubilized from the organic fraction than from the clays. However, although the results were highly reproducible, it was difficult to recognize a clear pattern for the remobilization of each metal. Similar studies were conducted on silicate-mineralized *B. subtilis* cells [85]. Wall-bound silicate proved to be very stable at low pH and in the presence of EDTA, Ca, and Na. Metals (Cd, Zn, Pb, Cu, and Ni) were most affected by acidity (pH 3.0), especially Pb and Cd. However, the efficiency of EDTA in extracting metals from the bacterial walls was lower than predicted from its high metal affinity constants, especially for Pb and Cd. Overall, it was clear that not only do bacteria compete better than clays for soluble metal, but that they also are less willing to exchange the metal when challenged by competing chemicals such as acids and organic chelating agents.

We believe that bacteria and their polymeric debris have a major impact on the mobility of toxic metals within natural ecological systems. Bacteria are ubiquitous and possess innate characteristics, such as their small size and the physicochemical nature of their surface, which make them ideal scavengers of metal ions and nucleators for mineral formation. Although there are other components of natural systems that are capable of binding metals, so far, bacteria seem to be among the most potent. As such, the presence of these small microorganisms likely represents one of the major cleansing mechanisms available to natural environments that are threatened with heavy metal contamination. Further studies, both in the laboratory and especially *in situ*, are needed if we are to completely understand the role of bacteria in metal immobilization and transport in natural systems and, perhaps, exploit this aspect of microbiology to deal with the problems of metal pollution that we have inherited from mining and industry.

Acknowledgments: All research from the authors' laboratory reported in this review has been supported by an on-going operating grant from the Natural Sciences and Engineering Research Council of Canada to T.J.B. M.U.M. was the recipient of a post-doctoral fellowship from the Spanish Ministry of Education and Science (1991-92) and is now supported by the same NSERC operating grant. S.S.-

L. was supported by a NSERC Graduate Scholarship while in T.J.B.'s laboratory.

REFERENCES

1.   Beveridge, T.J., "Wall Ultrastructure: How Little We Know," in *Antibiotic Inhibition of Bacterial Cell Surface Assembly and Function*, P. Actor, L. Daneo-Moore, M.L. Higgins, M.R.J. Salton, and G.D. Shockman, Eds., pp. 3-20 American Society for Microbiology, (Washington, DC, 1988).

2.   Davies, J.A., G.K. Anderson, T.J. Beveridge and H.C. Clark, "Chemical Mechanism of the Gram Stain and Synthesis of a New Electron-opaque Marker for Electron Microscopy Which Replaces the Iodine Mordant of the Stain," *J. Bacteriol.* 156:837-845 (1983).

3.   Beveridge, T.J. and J.A. Davies, "Cellular Responses of *Escherichia coli* and *Bacillus stubilis* to the Gram Stain," *J. Bacteriol.* 156:846-858 (1983).

4.   Beveridge, T.J., "Ultrastructure, Chemistry, and Function of the Bacterial Wall," *Int. Rev. Cytol.* 72:229-317 (1981).

5.   Beveridge, T.J. and L.L. Graham, "Surface Layers of Bacteria," *Microbiol. Rev.* 55:684-705 (1991).

6.   Rogers, H.J., J.B. Ward and I.D.J. Burdett, "Structure and Growth of the Cell Wall of Gram-positive Bacteria," *Symp. Soc. Gen. Microbiol.* 28:139-176 (1978).

7.   Beveridge, T.J., "Mechanisms of the Binding of Metallic Ions to Bacterial Walls and the Possible Impact on Microbial Ecology," in *Current Perspectives in Microbial Ecology,* M.J. Klug and C.A. Reddy, Eds., pp. 601-607 American Society for Microbiology, (Washington, DC, 1984).

8.   Beveridge, T.J. and W.S. Fyfe, "Metal Fixation by Bacterial Cell Walls," *Can. J. Earth Sci.* 22:1892-1898 (1985).

9.   Beveridge, T.J. and R.G.E. Murray, "Sites of Metal Deposition in the Cell Wall of *Bacillus subtilis*," *J. Bacteriol.* 141:876-887 (1980).

10. Doyle, R.J., T.H. Matthews and U.N. Streips, "Chemical Basis for Selectivity of Metal Ions by the *Bacillus subtilis* Cell Wall," *J. Bacteriol.* 143:471-480 (1980).

11. Beveridge, T.J., C.W. Forsberg and R.J. Doyle, "Major Sites of Metal Binding in *Bacillus licheniformis* Walls," *J. Bacteriol.* 150:1438-1448 (1982).

12. Beveridge, T.J. and R.G.E. Murray, "Uptake and Retention of Metals by Cell Walls of *Bacillus subtilis*," *J. Bacteriol.* 127:1502-1518 (1976).

13. Krell, P.J. and T.J. Beveridge, "The Structure of Bacteria and Molecular Biology of Viruses," *Int. Rev. Cytol.* 17:15-88 (1987).

14. Nikaido, H. and M. Vaara, "Outer Membrane," in *Escherichia coli and Salmonella typhimurium: Cellular and Molecular Biology, Volume One*, F.C. Neidhardt, J.L. Ingraham, K.B. Law, B. Megasanik, M. Schaechter and H.E. Umberger, Eds., pp. 7-23 American Society for Microbiology, (Washington, DC, 1987).

15. Beveridge, T.J., "The Structure of Bacteria," in *Bacteria in Nature, Volume 3*, J.S. Poindexter and E.R. Leadbetter, Eds., pp. 1-65 Plenum, (New York, 1989).

16. Hoyle, B.D. and T.J. Beveridge, "Metal Binding by the Peptidoglycan Sacculus of *Escherichia coli* K12," *Can. J. Microbiol.* 30:204-211 (1984).

17. Geesey, G.G. and L. Jang, "Interactions Between Metal Ions and Capsular Polymers," in *Metal Ions and Bacteria*, T.J. Beveridge and R.J. Doyle, Eds., pp. 325-358 John Wiley and Sons, (New York, 1989).

18. Messner, P., and U.B. Sleytr. "Crystalline Bacterial Cell-surface Layers," in *Advances in Microbial Physiology Vol. 33*, A.H. Rose and D.W. Tempest, Eds., pp. 213-275 Academic Press, (London, 1992).

19. König, H., "Archaeobacterial Cell Envelopes," *Can. J. Microbiol.* 34:395-406 (1988).

20. Beveridge, T.J., "Role of Cellular Design in Bacterial Metal Accumulation and Mineralization," *Ann. Rev. Microbiol.* 43:147-171 (1989).

21. Caldwell, D.E. and J.R. Lawrence, "Growth Kinetics of *Pseudomonas fluorescens* Microcolonies within the Hydrodynamic Boundary Layers of Surface Microenvironments," *Microb. Ecol.* 12:299-312 (1986).

22. Fletcher, M., "Effect of Solid Surfaces on the Activity of Attached Bacteria," in *Bacterial Adhesion: Physiological Implications*, D.C. Savage and M. Fletcher, Eds., pp. 339-362 Plenum, (New York, 1985).

23. Lion, L.W., M.L. Shuler, K.M. Hsieh and W.C. Ghiorse, "Trace Metal Interactions with Microbial Biofilms in Natural and Engineered Systems," *CRC Crit. Rev. Env. Cont.* 17:273-306 (1988).

24. Marshall, K.C., *Interfaces in Microbial Ecology* Harvard University Press, (Cambridge, 1976).

25. Fletcher, M., "The Attachment of Bacteria to Surfaces in Aquatic Environments," in *Adhesion of Microorganisms to Surfaces*, D.C. Ellwood, J. Melling and P. Rutter, Eds., pp. 87-108 Academic Press, (London, 1979).

26. Costerton, J.W., K.J. Cheng, G.G. Geesey, T.I. Ladd, J.C. Nickel, M. Dasgupta and T.J. Marrie, "Bacterial Biofilms in Nature and Disease," *Ann. Rev. Microbiol.* 41:435-464 (1987).

27. Neihof, R.A., and G.I. Loeb. "The Surface Charge of Particulate Matter in Seawater," *Limnol. Oceanogr.* 17:7-16 (1972).

28. Loeb, G.I. and R.A. Neihof, "Marine Conditioning Films," in *Advances in Chemistry, Series 145: Applied Chemistry at Protein Interfaces*, R.E. Baier, Ed., pp. 319-335 American Chemical Society, (Washington, DC, 1975).

29. Van Loosdrecht, M.C.M., J. Lyklema, W. Norde and A.J.B. Zehnder, "Influence of Surfaces on Microbial Activity," *Microbiol. Rev.* 54:75-87 (1990).

30. Van Loosdrecht, M.C.M. "Bacterial Adhesion: A Physico-chemical Approach," *Microb. Ecol.* 17:1-15 (1989).

31. Mills, A.L. and R. Maubrey, "Effect of Mineral Composition on Bacterial Attachment to Submerged Rock Surfaces," *Micro. Eco.* 7:315-322 (1981).

32. Marshall, K.C., "Mechanism of the Initial Events in the Sorption of Marine Bacteria to Surfaces," *J. Gen. Microbiol.* 68:337-348 (1971).

33. Marshall, K.C., "Bacterial Adhesion in Natural Environments," in *Microbial Adhesion to Surfaces*, R.C.W. Berkeley, J.M. Lynch, J. Melling, P.R. Rutter and B. Vincent, Eds., pp. 93-106 Ellis Horwood Ltd., (Chichester, 1980).

34. McCoy, W.F., J.D. Bryers, J. Robbins and J.W. Costerton, "Observations of Fouling Biofilm Formation," *Can. J. Microbiol.* 27:910-917 (1981).

35. Webster, J.J., G.J. Hampton, J.T. Wilson, W.C. Ghiorse and F.R. Leach, "Determination of Microbial Cell Numbers in Subsurface Samples," *Groundwater* 23:17-25 (1985).

36. Fliermans, C.B., "Microbial Life in the Terrestrial Subsurface of Southeastern Coastal Plain Sediments," *Haz. Water Haz. Materials* 6:155-171 (1989).

37. Beveridge, T.J., "The Bacterial Surface: General Considerations Towards Design and Function," *Can. J. Microbiol.* 34:363-372 (1988).

38. Beveridge, T.J. and S.F. Koval, "Binding of Metal Ions to Cell Envelopes of *Escherichia coli* K-12," *Appl. Env. Microbiol.* 42:325-335 (1981).

39. Ferris, F.G. and T.J. Beveridge, "Site Specificity of Metallic Ion Binding in *Escherichia coli* Lipopolysaccharide," *Can. J. Microbiol.* 32:52-55 (1986).

40. Faison, B.D., C.A. Cancel, S.N. Lewis and H.I. Adlers, "Binding of Dissolved Strontium by *Micrococcus luteus*," *Appl. Env. Microbiol.* 56:3649-3656 (1990).

41. Urrutia, M.M., M. Kemper, R. Doyle and T.J. Beveridge, "The Membrane-induced Proton-motive Force Influences the Metal-binding Ability of *Bacillus subtilis* Cell Walls," *Appl. Env. Microbiol.* 58:3837-3844 (1992).

42. Ferris, F.G., W.S. Fyfe and T.J. Beveridge, "Metallic Ion Binding by *Bacillus subtilis*: Implications for the Fossilization of Microorganisms," *Geology* 16:149-152 (1988).

43. Geesey, G.G., L. Jang, J.G. Jolley, M.R. Hankins, T. Iawoka and P.R. Griffiths, "Binding of Metal Ions by Extracellular Polymers of Biofilm Bacteria," *Water Sci. Technol.* 20:161-165 (1988).

44. Norberg, A.B. and H. Persson, "Accumulation of Heavy-metal Ions by *Zoogloea ramigera*," *Biotechnol. Bioeng.* 26:239-246 (1984).

45. McLean, R.J.C., D. Beauchemin, L. Clapham and T.J. Beveridge, "Metal-Binding Characteristics of the Gamma-Glutamyl Capsular Polymer of *Bacillus licheniformis* ATCC 9945," *Appl. Env. Microbiol.* 56:3671-3677 (1990).

46. Mayers, I.T. and T.J. Beveridge, "The Sorption of Metals to *Bacillus subtilis* Walls From Dilute Solutions and Simulated Hamilton Harbour (Lake Ontario) Water," *Can. J. Microbiol.* 35:764-770 (1989).

47. Mittelman, M.W. and G.G. Geesey, "Copper-binding Characteristics of Exopolymers from a Freshwater Sediment Bacterium," *Appl. Env. Microbiol.* 49:846-851 (1985).

48. Bitton, G. and V. Freihofer, "Influence of Extracellular Polysaccharide on the Toxicity of Copper and Cadmium Toward *Klebsiella aerogenes*," *Microb. Ecol.* 4:119-125 (1978).

49. Bremer, P.J. and M.W. Loutit, "The Effect of Cr(III) on the Form and Degradability of a Polysaccharide Produced by a Bacterium Isolated From a Marine Sediment," *Mar. Env. Res.* 20:249-259 (1986).

50. Couperwhite, I. and M.F. McCallum, "The Influence of EDTA on the Composition of Alginate Synthesized by *Azotobacter vinelandii*," *Arch. Microbiol.* 97:73-80 (1974).

51. Annison, G. and I. Couperwhite, "Consequences of the Association of Calcium with Alginate during Batch Culture of *Azotobacter vinelandii*," *Appl. Microbiol. Biotechnol.* 19:321-325 (1984).

52. Geddie, J.L. and J.W. Sutherland, "Uptake of Metals by Bacterial Polysaccharides," *J. Appl. Bacteriol.* 74:467-472 (1993).

53. Beveridge, T.J. and R.G.E. Murray, "Dependence of the Superficial Layers of *Spirillum putridiconchylium* on Ca^{2+}," *Can. J. Microbiol.* 22:1233-1244 (1976).

54. Kist, M.L. and R.G.E. Murray, "Components of the Regular Surface Array of *Aquaspirillum serpens* MW5 and Their Assembly *in vitro*," *J. Bacteriol.* 157:599-606 (1984).

55. Koval, S.F. and R.G.E. Murray, "Effect of Calcium on the *in vivo* Assembly of the Surface Protein of *Aquaspirillum serpens* VHA," *Can. J. Microbiol.* 31:261-267 (1985).

56. Beveridge, T.J., "Surface Arrays on the Wall of *Sporosarcina ureae*," *J. Bacteriol.* 139:1039-1048 (1979).

57. Beveridge, T.J., "Metal Ions and Bacteria," in *Metal Ions and Bacteria,* T.J. Beveridge and R.J. Doyle, Eds., pp. 1-30 John Wiley and Sons, (New York, 1989).

58. Sprott, G.D., T.J. Beveridge, G.B. Patel and G. Ferrante, "Sheath Disassembly in *Methanospirillum hungatei* Strain GP1," *Can. J. Microbiol.* 32:847-854 (1986).

59. Barghoorn, E.S. and S.A. Tyler, "Microorganisms from the Gunflint Chert," *Science* 147:563-577 (1965).

60. Beveridge, T.J., J.D. Meloche, W.S. Fyfe and R.G.E. Murray, "Diagenesis of Metals Chemically Complexed to Bacteria: Laboratory Formation of Metal Phosphates, Sulfides, and Organic Condensates in Artificial Sediments," *Appl. Env. Microbiol.* 45:1094-1108 (1983).

61. Ferris, F.G., T.J. Beveridge and W.S. Fyfe, "Iron-silica Crystallite Nucleation by Bacteria in a Geothermal Sediment," *Nature* 320:609-611 (1986).

62. Urrutia, M.M., and T.J. Beveridge. Unpublished.

63. Urrutia, M.M. and T.J. Beveridge, "Mechanism of Silicate Binding to the Bacterial Cell Wall in *Bacillus subtilis*," *J. Bacteriol.* 175:1936-1945 (1993).

64. Boogerd, F.C. and J.P.M. deVrind, "Manganese Oxidation by *Leptothrix discophora*," *J. Bacteriol.* 169:489-494 (1987).

65. Adams, L.F. and W.C. Ghiorse, "Characterization of Extracellular Mn^{2+} Oxidizing Activity and Isolation of a Mn^{2+}-oxidizing Protein From *Leptothrix discophora*," *J. Bacteriol.* 169:1279-1285 (1987).

66. Ferris, F.G., W.S. Fyfe and T.J. Beveridge, "Manganese Oxide Deposition in a Hot Spring Microbial Mat," *Geomicrobiol. J.* 5:33-41 (1987).

67. Chapnick, S.D., W.S. Moore and K.H. Nealson, "Microbially Mediated Manganese Oxidation in a Freshwater Lake," *Limnol. Oceanogr.* 26:1004-1014 (1982).

68. Ghiorse, W.C., "Biology of Iron- and Manganese-depositing Bacteria," *Ann. Rev. Microbiol.* 38:515-550 (1984).

69. Konhauser, K., unpublished.

70. Ferris, F.G., S. Schultze, T.C. Witten, W.S. Fyfe and T.J. Beveridge, "Metal Interactions with Microbial Biofilms in Acidic and Neutral pH Environments," *Appl. Env. Microbiol.* 55:1249-1257 (1989).

71. Ferris, F.G., K. Tazaki and W.S. Fyfe, "Iron Oxides in Acid Mine Drainage Environments and Their Association with Bacteria," *Chem. Geol.* 74:321-330 (1989).

72. Thompson, J.B., F.G. Ferris and D.A. Smith, "Geomicrobiology and Sedimentology of the Mixolimnion and Chemocline in Fayetteville Green Lake, New York," *Palaios* 5:52-75 (1990).

73. Brunskill, G.J. and S.D. Ludlam, "Fayetteville Green Lake, New York I: Physical and Chemical Limnology," *Limnol. Oceanogr.* 14:817-829 (1969).

74. Torgersen, T., D.E. Hammond, W.B. Clarke and T.-H. Peng, "Fayetteville Green Lake, New York: ^{3}H-^{3}He Water Mass Ages and Secondary Chemical Structure," *Limnol. Oceanogr.* 26:110-122 (1981).

75. Fogg, G.E., "Picoplankton," in *Perspectives in Microbial Ecology,* F. Mergusar and M. Gantar, Eds., pp. 96-100 Slovene Society for Microbiology, (Ljubljana, 1986).

76. Shively, J.M., and R.S. English. "The Carboxysome, a Prokaryotic Organelle," *Can. J. Bot.* 69:957-962 (1991).

77. Miller, A.G. and B. Colman, "Evidence for HCO_3 Transport by the Blue-green Alga (Cyanobacterium) *Coccochloris peniocystis*," *Plant Phys.* 65:397-402 (1980).

78. Thompson, J.B. and F.G. Ferris, "Cyanobacterial Precipitation of Gypsum, Calcite and Magnesite from Natural Alkaline Lake Water," *Geology* 18:995-998 (1990).

79. Schultze-Lam, S., G. Harauz and T.J. Beveridge, "Participation of a Cyanobacterial S Layer in Fine-grain Mineral Formation," *J. Bacteriol.* 174:7971-7981 (1992).

80. Ehrlich, H.L., and C.L. Brierley. "Bioleaching and Biobeneficiation," in *Microbial Mineral Recovery*, H.L. Ehrlich and C.L. Brierley, Eds., pp. 1-182 McGraw-Hill, (New York, 1990).

81. Walker, S.G., C.A. Flemming, F.G. Ferris, T.J. Beveridge and G.W. Bailey, "Physicochemical Interaction of *Escherichia coli* Cell Envelopes and *Bacillus subtilis* Cell Walls with Two Clays and Ability of the Composites to Immobilize Metals From Solution," *Appl. Env. Microbiol.* 55:2976-2984 (1989).

82. Carroll, D., "Role of Clay Minerals in the Transport of Iron," *Geochim. Cosmochim. Acta* 14:1-27 (1958).

83. Pickering, W.F., "Copper Retention by Soil/Sediment Components," in *Copper in the Environment, Part I: Ecological Cycling*, J.O. Nriagu, Ed., pp. 217-253 John Wiley and Sons, (Toronto, 1979).

84. Flemming, C.A., F.G. Ferris, T.J. Beveridge and G.W. Bailey. "Remobilization of Toxic Heavy Metals Adsorbed to Bacterial Wall-Clay Composites," *Appl. Env. Microbiol.* 56:3191-3203 (1990).

85. Urrutia, M.M. and T.J. Beveridge, "Remobilization of Heavy Metals Retained as Oxyhydroxides or Silicates by *Bacillus subtilis* Cells," *Appl. Env. Microbiol.* 59:43323-4329 (1993).

6

DETERMINATION OF REDOX STATUS IN SEDIMENTS

Tim Grundl
Department of Geosciences
University of Wisconsin - Milwaukee
Milwaukee, WI 53201

1. INTRODUCTION

The redox status of natural sediment-water systems is one of the major factors influencing the speciation and mobility of metals. The redox status of a system can be described in terms of an intensity factor (redox potential) or a capacity factor (poising capacity). Redox status is one of the most difficult parameters to measure in natural systems. The fundamental problem lies in the fact that both intensity and capacity factors are thermodynamically defined and are valid only at equilibrium, whereas natural systems are seldom at redox equilibrium.

This chapter will present the underlying thermodynamic principles inherent to redox measurements (both intensity and capacity) in ideal, equilibrium systems and compare this to redox measurements taken in real, disequilibrium systems. A discussion of alternate methods for assessing redox status that do not necessitate solution equilibrium will also be presented.

2. REDOX INTENSITY IN IDEAL SYSTEMS

The redox intensity is usually measured by direct potentiometry between a reference and an inert (typically Pt) electrode. The measured potential is reported as mV versus the standard hydrogen electrode (Eh) or as the negative logarithm of the electron activity (pE). The relation between the two is defined by:

$$Eh = 59.2 \ (pE) \tag{1}$$

Direct potentiometry measures the potential adopted by the inert electrode when the net current across the electrode/solution interface is zero. The electrode responds to a summation of all electron exchanges across the interface and adopts a potential such that the anodic current (arising from the oxidation of reduced species) exactly balances the cathodic current (arising from the reduction of oxidized species). The equation that defines the current potential relationship at an electrode surface is the Butler-Volmer equation. For a single redox couple, the Butler-Volmer equation is written:

$$i = \left\{ F(k_f^e)M_{ox} \exp\left[\frac{nF}{2RT}(E)\right] \right\} - \left\{ F(k_r^e)M_{red} \exp\left[\frac{-nF}{2RT}(E)\right] \right\} \tag{2}$$

where:

k_f^e = intrinsic rate constant for the passage of electrons from the electrode to solution

k_r^e = intrinsic rate constant for the passage of electrons to the electrode from solution

E = potential between electrode and solution

i = net current across electrode interface

n = number of electrons exchanged

M_{ox} = molarity of the oxidized form of species M

M_{red} = molarity of the reduced form of species M

F = 23,062 cal/volt - equiv

R = 1.987 cal/°K - mole

T = temperature (°K).

Equation 2 is general in nature and defines the current-potential relation for any situation. The first term on the right side represents the cathodic current and the second term represents the anodic current. The potential (E) can be split into two components:

$$E = E_{eq} + \Delta E \tag{3}$$

where:

E_{eq} = potential at equilibrium between electrode and solution

ΔE = overpotential

Equation 2 can then be written:

$$i = \left\{ F(k_f^e) M_{ox} \exp\left[\frac{nF}{2RT}(E_{eq})\right]\right\} \exp\left[\frac{nF}{2RT}(\Delta E)\right] - $$
$$\left\{ F(k_f^e) M_{red} \exp\left[\frac{-nF}{2RT}(E_{eq})\right]\right\} \exp\left[\frac{-nF}{2RT}(\Delta E)\right] \qquad (4)$$

For the special case at equilibrium, $\Delta E = 0$, the anodic and cathodic currents are equal and there is no net current across the interface (i = 0). Note that although the net current is zero at equilibrium, there is still electron exchange occurring in both the anodic and cathodic directions. The terms inside the braces in Equation 4 represent the equilibrium exchange current and are denoted by i_0. Equation 4 can then be rewritten:

$$i = i_0 \left\{ \exp\left[\frac{nF}{2RT}(\Delta E)\right] - \exp\left[\frac{-nF}{2RT}(\Delta E)\right]\right\} \qquad (5)$$

Figure 1 is a plot of the current versus overpotential for a single redox couple as described by Equation 4.

Equilibrium exchange currents (i_0) are a function of the concentration of redox species present (M_{ox} and M_{red}) and the intrinsic rate of electron exchange k_f^e and k_r^e. The net current across the interface (i) is a function not only of the concentration and electron exchange rate of the redox couple, but also of the distance from equilibrium (ΔE).

In ideal systems the electrode interface is at equilibrium and by setting both i and ΔE to zero, Equation 4 can be expressed as:

$$F(k_f^e) M_{ox} \exp\left[\frac{nF}{2RT}(E_{eq})\right] = F(k_f^e) M_{red} \exp\left[\frac{-nF}{2RT}(E_{eq})\right] \qquad (6)$$

Rearrangement yields

$$\exp\left[\frac{nF}{RT}(E_{eq})\right] = \frac{(k_f^e)M_{ox}}{(k_f^e)M_{red}} \tag{7}$$

$$E_{eq} = \frac{RT}{nF}\ln\left(\frac{k_r^e}{k_f^e}\right) + \frac{RT}{nF}\ln\left(\frac{M_{ox}}{M_{red}}\right) \tag{8}$$

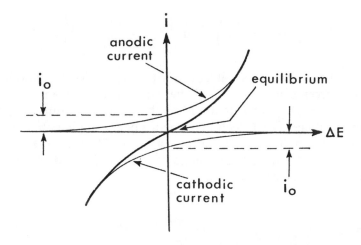

Figure 1.   ***Plot of the Current at the Electrode-solution Interface Versus Overpotential for a Single Redox Couple. The net current across the interface is the sum of the anodic and cathodic currents.***

The first term on the right side of Equation 8 is the formal potential for the redox couple of interest. If the concentration terms (M_{ox} and M_{red}) are expressed as activities and E_{eq} is expressed relative to the standard hydrogen electrode, Equation 8 becomes the familiar Nernst equation:

$$E_{eq} = E° + \frac{RT}{nF}\ln\left(\frac{A_{ox}}{A_{red}}\right) \tag{9}$$

where:

$E°$ =   standard potential of the redox couple

A_{ox} =   activity of the oxidized species

A_{red} = activity of the reduced species

This electrokinetic approach to the Nernst equation differs from the usual thermodynamic approach using free energies of reaction. This approach demonstrates that the measured potential in a system at equilibrium is a thermodynamically valid measurement and as such would serve as a useful master variable.

3.   REDOX INTENSITY IN DISEQUILIBRIUM SYSTEMS

The preceding discussion holds true only if the system behaves ideally. If a potential measurement is to represent the thermodynamically defined redox intensity within a system, that system must conform to the following three conditions:

1)   the solution must be at internal, homogeneous equilibrium,

2)   the system solution must be at heterogeneous equilibrium with any solids present, and

3)   electrochemical equilibrium must exist at the electrode/solution interface.

If any of these conditions are not met, the system is at redox disequilibrium and the potential adopted by an electrode is a mixed potential in which the anodic current from one couple is balanced by the cathodic current from one or more of the other redox couples present. Each redox couple exchanges electrons independently and the electrode responds to a summation of independent anodic and cathodic currents. The resultant mixed potential can be defined in a system with n redox couples by:

$$0 = \sum_{j=i}^{n} i_{oj} \left\{ \exp\left[\frac{n_j F(E_{mix} - E_{eqj})}{2RT} \right] - \exp\left[\frac{n_j F(E_{mix} - E_{eqj})}{2RT} \right] \right\} \quad (10)$$

where:

E_{mix} = the mixed potential

E_{eqj} = the potential at electrode/solution equilibrium for couple j.

The mixed potential (E_{mix}) is adopted by the electrode such that the net current across the electrode surface is zero. Note that the E_{mix} is at equilibrium with no single couple. The overpotential terms (E_{mix}-E_{eqj}) represent the extent of disequilibrium between E_{mix} and each couple present. The potential adopted by the electrode (E_{mix}) is an arbitrary mix of individual exchange currents and is not definable in thermodynamic terms because it requires kinetic information on the electron exchange rate of each couple as embodied in i_{oj}.

An additional problem arises from the slow kinetics of many redox reactions which cause each E_{eqj} to change through time as each couple approaches equilibrium. These changing E_{eq} values cause a corresponding time variance in the potential adopted by the electrode (E_{mix}). Therefore a full description of E_{mix} also requires kinetic information on each redox couple in solution. Detailed kinetic information, both in bulk solution and at the electrode interface, is not readily available and E_{mix} is undefinable in practical terms.

Peiffer *et al.* [1] have defined sensor effective redox couples (SERC) and sensor ineffective redox couples (SIRC) based on the relative contribution each couple makes to the mixed potential measured at an electrode. SERCs contribute sufficient exchange current to measurably affect electrodes measurements whereas SIRCs do not affect electrode measurements. Inorganic SERCs are limited to $Fe^{3+}/Fe^{2+}, S^0/S^{2-}$, Mn^{4+}/Mn^{2+} and O_2/H_2O_2. Organic SERCs are largely limited to quinone/hydroquinone, although disulfide bridges (as in cystine) or organic complexes with iron may be important contributors to mixed potentials [1].

4. REDOX INTENSITY IN NATURAL SYSTEMS

In order to determine the usefulness of measured redox potential in natural systems, these systems must be evaluated in terms of the three conditions necessary for ideal systems. Widespread homogeneous and heterogeneous redox disequilibria in natural systems is well documented [2-5]. Lindberg and Runnells [3] examined over 600 high quality water analyses and compared multiple redox potentials calculated from analytically determined concentrations of redox couples and found deviations as large as 1000 mV. This large deviation between couples in the same solution is indicative of homogeneous disequilibrium. Comparison between any single calculated redox potential and measured potential showed discrepancies as large as 500 mV [3]. Similar discrepancies between different calculated potentials and between calculated potentials and those measured at platinum electrodes were reported by Barcelona *et al.* [4] and Holm and Curtis [5]. However, Barcelona *et al.* [4] report general agreement between Pt measurements and the O_2/H_2O_2 couple in aerobic water and Fe^{3+}/Fe^{2+} in anaerobic water when averaged over a two year period.

This disequilibrium is the result of two factors, the first of which is the slow kinetics of most environmentally important redox reactions. The second factor is the effect of photosynthetic life forms. These life forms are in essence thermodynamically unstable compounds that use the steady influx of solar energy to reduce CO_2 gas to organic matter [7]. After death, the oxidation of these organic compounds back to CO_2 is accomplished by non-photosynthetic microbes using a variety of electron acceptors (oxidants). The smaller energy yield for each successive oxidant gives rise to specific environments in which organic matter is being oxidized by a specific oxidant. The redox intensity in each environment is dominanted by a single redox couple. Stumm [7] lists the important naturally occurring microbially mediated redox reactions. The Eh of the system will lie close to the formal potential of the dominant redox couple. Table 1 lists the formal potentials of several environmentally important redox couples [9]. The formal potentials are calculated for pH = 7 and partial pressure of gases set to 1 atm. For solid phases, concentrations are the average number of moles of solid surface per volume of water [9].

Table 1. Environmentally Important Redox Couples in Natural Waters.

Reaction	$E°_{pH=7}(V)$
$O_2(g) + 4H^+ + 4e^- -> 2H_2O$	0.811
$NO_3^- + 10H^+ + 8e^- -> NH_4^+ + 3H_2O$	0.323
$MnO_2(s) + 4H^+ + 2e^- -> Mn^{2+} + 2H_2O$	0.379
$Fe(OH)_3(s) + 3H^+ + e^- -> Fe^{3+} + 3H_2O$	-0.295
$SO_4^{2-} + 9H^+ + 8e^- -> HS^- + 4H_2O$	-0.221
$CO_2(g) + 4H^+ + 4e^- -> CH_2O + H_2O$	-0.484

The third condition necessary for an ideal system, equilibrium between electrodes and solution, is largely an operational problem and is a function of the electrode surface and its pre-treatment [2,6-8]. The most common cause of electrochemical disequilibrium is electrode poisoning. Electrode poisoning is the result of alteration of the electrode surface, usually by the adsorption of material onto the surface. This alters or diminishes the exchange current at the electrode surface and prevents the electrode from attaining equilibrium with the surrounding solution. Electrode poisoning can be particularly serious with platinum electrodes. Platinum often does not behave as if it were inert but reacts selectively with oxygen to form a surface oxide layer [8,10,11]. Platinum can also adsorb substances such as sulfide, cyanide, and organic matter which form surface films [8,11]. Comprehensive reviews on the use of electrodes in geochemical studies have been published [12-14]. Electrode materials and pretreatment techniques designed to minimize poisoning are discussed by several authors [8,10,11,15-21].

5.   ALTERNATE TECHNIQUES TO MEASURE REDOX INTENSITY

Some of the difficulties with traditional potentiometric determinations of redox potential can be circumvented with modified measurement techniques. Drifting or unstable potentials are often obtained in poorly poised natural systems because exchange currents are insufficient. Under ideal conditions, modern potentiometers need roughly 10^{-7} A of both anodic and cathodic exchange current to affect a measurement. For the highly reversible couple Fe^{3+}/Fe^{2+}, this corresponds to a concentration of approximately 10^{-6} M [21]. Liu and Yu [22] have proposed measuring redox potential in poorly poised systems by deliberately polarizing the working electrode and recording the depolarization on a log-time scale. The resulting linear plots, for several degrees of polarization, converge at a limiting sample potential. This technique gives a stable potential by inducing large exchange currents during electrode polarization. However, the resulting potential is still a mixed potential that is dependent on the concentration and lability of all redox couples present.

An alternative approach to measuring the redox potential of a solution employs redox indicators [23,24]. The indicators form labile and reversible redox couples with differently colored conjugates. A suite of indicators with a range of formal potentials is selected and a very small quantity of each is added to an aliquot of sample. Assuming the indicator equilibrates with the sample and the sample was well poised with respect to the amount of indicator added, the sample redox potential is approximately the formal potential of the dye that divides those dyes that were visibly reduced and those that were not. Obviously this technique is not appropriate for strongly colored samples or soil/sediment slurries, but the indicators may be useful as "mediators" of redox reactions between the sample and an electrode.

A mediation technique specifically designed for soils has been proposed by Lamm [25]. Small aliquots of the sample are treated with mediator solutions containing various proportions of $K_3Fe(CN)_6$ and $K_4Fe(CN)_6$. The change in potential measured at an electrode after the addition of sample is then plotted against the calculated potential of the pure mediator solution. The Eh of the soil is the potential of the solution that exhibits no change upon the addition of sample.

Breck [26] proposed that redox mediators be used to make all redox potential measurements specific to the same chemical-electrode interactions. If a mediator (in a very small quantity relative to the redox poising capacity of the sample) is added to the sample and allowed to equilibrate, the redox potential of the mediator can then be measured with an electrode, spectrophotometer, or polarograph. Breck [26] used $Fe(CN)_6^{3-}/Fe(CN)_6^{4-}$ and I_2/I as mediators and a Pt electrode for detection in seawater samples. Cherry et $al.$ [27] proposed the use of naturally occurring As^{5+}/As^{3+} as a mediator to determine the redox state of natural solutions. Bisogni [28] used Hg^{2+}/Hg° as a mediator to assess the redox potential exerted by O_2/H_2O couple in oxygenated water.

All these methods use the mediator couple as the sensing element instead of an electrode. The necessity for electrochemical equilibrium between sample and electrode is replaced by the necessity for chemical equilibrium between mediator and sample. Internal redox equilibrium must exist within the solution if a mediator defined potential can be considered a Nernstian potential. Thus mediator defined potentials are not inherently better than electrode defined potentials at providing system-wide redox conditions.

6.   REDOX CAPACITY IN IDEAL SYSTEMS

The poising capacity (ρ) is defined in terms of the amount of strong oxidant needed to change the redox potential of the system. This can be written as the differential ratio:

$$\rho = \frac{dO_{tot}}{dEh} \tag{11}$$

where O_{tot} = total concentration of strong oxidant.

In ideal systems in which redox equilibrium is maintained, the poising capacity is unique and its characteristics can be ascertained. Nightingale [29] evaluated the poising capacity of ideal systems using the Nernst equation (Equation 9) in conjunction with mass balance and charge balance equations. Grundl [30] discusses poising capacity in disequilibrium systems.

If a solution containing m redox couples is titrated with a strong oxidant, the poising capacity, ρ, is defined [29]:

$$\rho = \frac{\sum_{j=i}^{m}\left[\frac{(n_j)^2(C_j)F}{RT}\operatorname{sech}^2\left(\frac{\gamma_j}{2}\right)\right] + \frac{(n_o)^2(O_{tot})F}{RT}\operatorname{sech}^2\left(\frac{\theta}{2}\right)}{-n_o\left[\tanh\left(\frac{\theta}{2}\right)-1\right]} \qquad (12)$$

where:

C_j = total concentration of redox couple j

n_j = number of electrons transferred by redox couple $\gamma_j = \frac{n_jF}{RT}(E_{eq}-E_j^\circ)$

n_o = number of electrons transferred by the oxidant

m = number of redox couples present

For a system at internal redox equilibrium, E_{eq} represents the thermodynamic potential or Eh of the system as defined by the Nernst equation. Figure 2 is a plot of poising capacity versus potential for a solution containing 1 mM concentrations of two redox couples. The formal potentials (i.e., the potential at equimolar concentrations of the oxidized and reduced species) of these couples are -0.20 V and +0.40 V and one electron is exchanged in each case. The oxidant has a formal potential 1.00 V. The poising capacity reaches a rather sharp maximum at each formal potential where there are significant quantities of both oxidized and reduced species present in solution. The height of each peak is proportional to the total concentration of redox species present and to the square of the number of electrons transferred. The width of each peak is only slightly affected by these variables. The rise in poising capacity near 1.00 V is due to the presence of strong oxidant.

Since the solution is at equilibrium, the concentration ratios of each couple are unique at a given potential. For example at a potential of +0.40 V, one couple is half oxidized and half reduced while the other couple is completely oxidized. At a potential of -0.40 V, both couples are completely reduced. The potential adopted by an electrode in this solution (Eh) is shown in Figure 2. At this potential one couple is essentially all reduced and the other couple is essentially all oxidized.

There is very little exchange current and the solution is poorly poised. Figure 3 is calculated for the same conditions except the formal potentials are set at +0.40 V and +0.50 V. In this case the individual poising capacities overlap and the Eh adopted by the electrode is well poised.

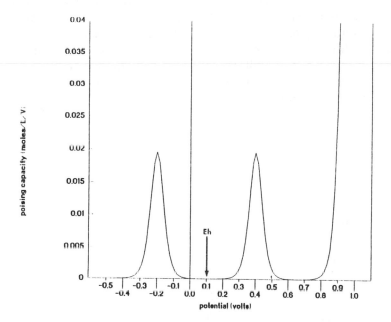

Figure 2.   *Poising Capacity of an Ideal Solution Containing 1 mM Concentrations of Two Single Electron Couples with Potentials of -0.20 V and +0.40 V. Solution was titrated with an oxidant with a formal potential of +1.00 V. The Eh of this solution is poorly poised.*

7.   REDOX CAPACITY IN DISEQUILIBRIUM SYSTEMS

The measurement of the true poising capacity of a real system is predicated upon the ability to measure the true redox potential. In systems typified by redox disequilibria, the measured potential (E_{mix}) is governed by Equation 10. Each couple independently exchanges electrons across the electrode surface and concentration ratios of each couple in solution are not uniquely defined by the measured potential. Additionally, each couple reacts with the oxidant in an independent manner (compare with the ideal case). The addition of an oxidant causes each individual E_{eq} to change through time to a new value according to the pertinent kinetics for each couple. Note that in most real systems, each couple reacts with the oxidant to a different extent and has a different final E_{eqj} (i.e., system disequilibrium). The overpotential terms in Equation 10 ($E_{mix}-E_{eq}$) are controlled by both bulk solution kinetics and extent of disequilibria. In order to maintain the net current at zero, the E_{mix} adopted by the electrode will continually vary until all E_{eqj} terms remain constant. If sufficient time is allowed after the addition of oxidant for all redox reactions to reach completion, a new steady state E_{mix} will be obtained. A value for poising capacity can be calculated from this E_{mix}. However, it is not indicative of equilibrium conditions within the system ($dO_{tot}/dE_{mix} \neq dO_{tot}/dEh$). The time dependence for the attainment of the final poising capacity and the lack of equilibrium at the final state both conspire to prevent measured ρ values from having any thermodynamic significance.

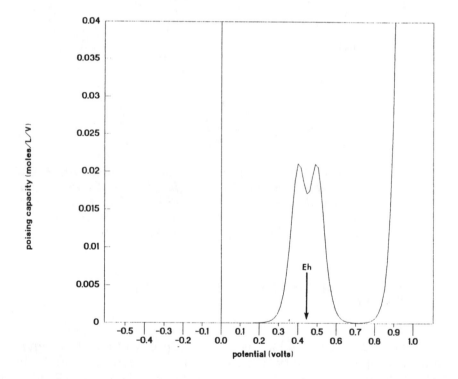

Figure 3.   *Poising Capacity of an Ideal Solution Containing 1 mM Concentrations of Two Single Electron Couples with Potentials of -0.40 V and +0.50 V. Solution was titrated with an oxidant with a formal potential of +1.00 V. The Eh of this solution is well poised.*

8.   REDOX CAPACITY IN NATURAL SYSTEMS

Several authors have measured the apparent poising capacity in real systems by titration with any of a variety of oxidants. Barcelona and Holm [31] measured the apparent poising capacity of aquifer solids in the oxidative direction with both dichromate and hydrogen peroxide and in the reductive direction with chromous ion (CrII). Frevert [32] titrated lake sediments with oxygen gas and defined an operational poising capacity from the resulting platinum electrode response. Patrick [33] studied apparent redox capacity in flooded paddy soils and related it to a variety of inorganic reducible species. Tian-Ren [34] suggested using the depolarization rate of platinum electrodes to measure redox capacity in sediment-water systems. This technique is equivalent to "titrating" the system with metallic electrons. In all cases the measured capacities yield useful information on the stability of the individual system to changes in redox potential, however, interpretation of these results on the basis of thermodynamic principles is unwarranted.

An additional complication arises in the interpretation of redox capacity measurements if the coupling between poising and buffering is considered. All environmentally important redox reactions have an acid base character either directly or indirectly (Table 1). Any buffer capacity will indirectly contribute to the poising capacity of the solution.

The presence of solids in the system contributes significantly to both the measured poising and buffering capacity. Barcelona and Holm [31] report less than 1 percent of the poising capacity of an organic-poor sand and gravel aquifer system is attributable to dissolved species in the groundwater. Aquifer solids contribute up to 0.41 meq/g oxidative capacity and 0.14 meq/g reductive capacity [31]. The solid phase poising capacity may arise from direct electron transfer from solid to aqueous species or alternatively the solids may serve as a nearly infinite source (or sink) of dissolved solutes to the groundwater. It is likely that microbes mediate either of these processes. Further research is necessary to determine which of these mechanisms is operative.

9.   ALTERNATE METHODS OF DESCRIBING REDOX STATUS

It is clear from the preceding discussion that thermodynamically valid Nernstian potentials are not definable in natural aqueous systems. With this in mind, several authors have proposed alternate methods of describing redox status of natural systems that are not tied to strict thermodynamic principles. These techniques all abandon the idea of a single, system-wide potential in favor of describing the system in terms of which redox couple is dominant.

Redox conditions in natural aqueous systems are typically controlled by the microbial oxidation of organic carbon. The two most important electron acceptors are oxygen and sulfur. Berner [35] made use of this fact as the basis for his redox classification scheme. This scheme divides aqueous systems into five categories based on the sequential application of three qualitative tests (Figure 4). The presence of oxygen indicates little organic carbon was present to start with and the system remains oxic. If sufficient organic carbon is present to exhaust the supply of oxygen, the system becomes anoxic. If an oxidized form of sulfur is present (S° or SO_4^{2-}), the oxidation of organic carbon proceeds via the reduction of sulfur to sulfide and the system becomes sulfidic. If organic carbon still remains, the system reverts to fermentation reactions and becomes methanic. Post oxic systems contain enough carbon to cause anoxia but insufficient to reduce all the oxidized sulfur. This classification system is analogous to the classification of lakes by trophic level. Both systems are qualitative, easy to apply and yield much useful information on the prevalent conditions in each system.

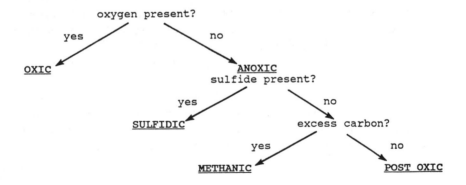

Figure 4.   *The Geochemical Classification Scheme as Proposed by Berner [35].*

Frevert [32] suggested the use of a controlled feedback system to measure the response of a platinum electrode to individual redox couples in a H_2S bearing natural sediment/water system. The system was titrated with dissolved oxygen gas (DO) which served as the electron acceptor. Electron donors were unknown but were distinguished on the basis of changes in the $\Delta E_{mix}/\Delta DO$. As the titration continued, the measured potential responded to specific redox couples. This technique does not demand internal redox equilibrium, but instead empirically relates sensor response $\Delta E_{mix}/\Delta DO$) to solution conditions.

A third way to define redox status is to calculate a potential from analytically determined concentrations of redox active species. The reaction or process of interest can then be related to the redox couple that is serving as the primary electron shuttle.

Alternate methods of describing redox capacity have also been suggested. Scott and Morgan [9] and Barcelona and Holm [31] use a capacity measurement that does not require information on bulk solution kinetics or on electron exchange kinetics at the electrode surface. This measurement is called the total oxidative capacity (OXC) and presumes complete reaction between all redox couples present. An arbitrary division between those couples considered oxidants and those considered reductants must be made and the stoichiometric sum of the reductants present is subtracted from the stoichiometric sum in the oxidants present. The difference is the OXC and represents the total inherent oxidative (or reductive) capacity of the system if it is fully reacted. This is equivalent to assuming redox equilibrium. This capacity measurement is a useful one in that it is a conservative measurement that depends solely on the concentrations of redox couples present but it only represents the limiting case at equilibrium and does not represent actual conditions within the solution.

10.   SUMMARY

The description of the redox intensity of natural systems is constrained by theoretical and practical limitations that typically prevent a correlation between measured potentials and a thermodynamically defined Nernstian potential. This is particularly true in reduced systems where no single redox couple dominates. The primary limitation is that natural systems seldom achieve either internal, homogeneous redox equilibrium or heterogeneous equilibrium with redox active solid species. Methods other than direct potentiometry at an inert electrode have been proposed to measure redox potentials in natural systems including the depolarization of a pre-polarized working electrode, use of electron exchange mediators and redox sensitive dyes. The underlying problem of redox disequilibria in natural systems precludes any of these techniques from yielding a Nernstian Eh. Widespread redox disequilibria also prevents the measurement of thermodynamically defined poising capacities.

One solution to this dilemma is to abandon the idea of a single, thermodynamically defined potential in favor of multiple, solute specific redox potentials. These potentials are defined in terms of specific redox couples and although they do not represent system wide conditions, solute specific potentials are thermodynamically valid within the context of each redox couple. Solute specific potentials will give an indication of the redox conditions relevant to reactions that are mediated by specific redox couples. Possible techniques for the determination of solute specific redox potentials include the use of aqueous phase mediators that selectively respond to specific redox couples via the use of calculated potentials derived from analytically determined concentrations of redox active species. A second solution to this dilemma is to qualitatively describe redox status by methods such as Berner's geochemical classification scheme [35], or the total oxidative capacity of Scott and Morgan [9].

REFERENCES

1.   Peiffer, S., O. Klemm, K. Pecher and R. Hollerung, "Redox Measurements in Aqueous Solutions - A Theoretical Approach to Data Interpretation Based on Electrode Kinetics," *J. Contam. Hydrol.* 10:1-18 (1992).

2.   Morris, J. and W. Stumm, "Redox Equilibria and Measurements of Potentials in the Aquatic Environment," *ACS Advances in Chemistry* Series, 67:270-285 (1967).

3.   Lindberg, R. and D. Runnells, "Ground Water Redox Reactions: An Analysis of Equilibrium State Applied to Eh Measurements and Geochemical Modeling," *Science* 225:925-927 (1984).

4.   Barcelona, M.J., T.R. Holm, M.R. Schock and G.K. George, "Spatial and Temporal Gradients in Aquifer Oxidation-reduction Conditions, "*Water Resources Res.* 25:991-1003 (1989).

5.   Holm, T.R. and C.D. Curtis, "A Comparison of Oxidation-reduction Potentials Calculated From the As(V)/As(III) and Fe(III)/Fe(II) Couples With Measured Platinum Electrode Potentials in Groundwater," *J. Contam. Hydrol.* 5:67-81 (1989).

6.   Hostettler, J., "Electrode Electrons, Aqueous Electrons, and Redox Potentials in Natural Waters," *Amer. J. Science* 284:734-759 (1984).

7.   Stumm, W., "Redox Potential as an Environmental Parameter; Conceptual Significance and Operational Limitation," *Proc. Third International Conference of Water Pollution Research, Munich, Germany*, 1:283-308 (1966).

8.   Whitfield, M., "Thermodynamic Limitations on the Use of the Pt Electrode in Eh Measurements," *Limnol. Oceanogr.*, 19:857-865 (1974).

9.   Scott, M. and J. Morgan, "Energetics and Conservative Properties of Redox Systems," in *Chemical Modeling of Aqueous Systems II*, D.C. Melchior and R.L. Bassett, Eds., pp. 368-378 ACS Symposium Series 416, (Washington, 1990).

10. Solomin, G.A., "Methods of Determining Eh and pH in Sedimentary Rocks," pp. 33-56 Consultant Bureau, (New York, 1965).

11. Vershinin, A.V. and A.G. Rozanov, "The Platinum Electrode as an Indicator of Redox Environment in Marine Sediments," *Marine Chem.* 14:1-15 (1983).

12. Back, W. and I. Barnes, "Equipment for Field Measurement of Electrochemical Potentials," *U.S. Geol. Survey Prof. Paper* 280-C:366-368 (1961).

13. Back, W. and I. Barnes, "Relation of Electrochemical Potentials and Iron Content to Groundwater Flow Patterns," *U.S. Geol. Survey Prof. Paper* 498-C:1-16 (1965).

14. Langmuir, D., "Eh-pH Determinations," in *Procedures in Sedimentary Petrology*, R.E. Carver, Ed. pp. 597-635 Wiley Interscience, (New York, 1971).

15. Ives, J.D. and G.J. Janz, "Reference Electrodes, Theory and Practice," p. 651 Academic Press, (New York, 1961).

16. Engstrom, R.C., "Electrochemical Pretreatment of Glassy Carbon Electrodes," *Anal. Chem.* 54:2310-2314 (1982).

17. Hughes, S. and D.C. Johnson, "High-performance Liquid Chromatographic Separation With Triple-pulse Amperometric Detection of Carbohydrates in Beverages" *J. Agric. Food Chem.* 30:712-714 (1982).

18. Austin, D.S., J.A. Polta, A.P, Polta, C. Tang, T.D. Cabelka and D. Johnson, "Electrocatalyses at Platinum Electrodes for Anodic Electroanalysis," *J. Electroanal. Chem.* 168:227-248 (1984).

19. Bohn, H.L., "Electromotive Force of Inert Electrodes in Soil Suspensions" *Soil Sci. Soc. Amer. Proc.* 32:211-215 (1968).

20. Shaikh, A.U., R.M. Hawk, R.A. Sims and H.D. Scott, "Graphite Electrode for the Measurement of Redox Potential and Oxygen Diffusion Rate in Soils," *Nuc. Chem. Waste Mgt.* 5:237-243 (1985).

21. Stumm, W. and J. Morgan, *Aquatic Chemistry* John Wiley and Sons, (New York, 1981).

22. Liu, Z. and T. Yu, "Depolarization of a Platinum Electrode in Soils and its Utilization for the Measurement of Redox Potential," *J. Soil Science*, 35:469-479 (1984).

23. Jacob, H.E., "Redox Potential," in *Methods in Microbiology*, J.R. Norris, and D.W. Ribbons, Eds. pp. 92-123 Academic Press, (London, 1970).

24. Zobell, C.W., "Studies on Redox Potential of Marine Sediments," *AAPG Bull.* 30:477-513 (1946).

25. Lamm, C.G., "Reduction-oxidation Level of Soils," *Nature* 177:620-621 (1956).

26. Breck, W.G., "Redox Potentials by Equilibration," *J. Marine Res.* 30:121-139 (1972).

27. Cherry, J.A., A.U. Saikh, D.E. Tallman and R.V. Nicholson, "Arsenic Species as an Indicator of Redox Conditions in Groundwater," *J. Hydrol.* 43:373-392 (1979).

28. Bisogni, J.J., "Using Mercury Volatility to Measure Redox Potential in Oxic Aqueous Systems," *Environ. Sci. Technol.* 23:828-831 (1989).

29. Nightingale, E. "Poised Oxidation Reduction Systems," *Anal. Chem.* 30:267-272 (1958).

30. Grundl, T., "A Review of the Current Understanding of Redox Capacity in Natural, Disequilibrium Systems," *Chemosphere* 28:613-626 (1994).

31. Barcelona, M. and T. Holm, "Oxidation-reduction Capacities of Aquifer Solids," *Environ. Sci. Technol.* 25:1565-72 (1992).

32. Frevert, T., "Can the Redox Conditions in Natural Waters be Predicted by a Single Parameter?," *Schweiz. Z. Hydrol.* 46:269-290 (1984).

33. Patrick, W.H., "The Role of Inorganic Redox Systems in Controlling Reduction in Paddy Soil," in *Proc. Symp. on Paddy Soils*, pp. 107-117 Springer-Verlag, (Berlin, 1981).

34. Tian-Ren, Y., "Oxidation-reduction Properties of Paddy Soils," in *Proc. Symp. on Paddy Soils*, pp. 95-106 Springer-Verlag, (Berlin, 1981).

35. Berner, R.A., "A New Geochemical Classification of Sedimentary Environment," *J. Sed. Pet.* 51:359-365 (1981).

7

CHANGES IN METAL SPECIATION FOLLOWING ALTERATION OF SEDIMENT REDOX STATUS

John H. Pardue and William H. Patrick, Jr.
Wetland Biogeochemistry Institute
Center for Coastal, Energy and Environmental Resources
Louisiana State University
Baton Rouge, LA 70803

1. INTRODUCTION

Changes in oxidation-reduction (redox) potential can result from natural (e.g., flooding) and anthropogenic (e.g., dredging) alterations of sediment conditions. Changes in metal and metalloid speciation generally follow these changes in sediment Eh and pH conditions. Alteration of sediment Eh and pH conditions are often driven by microbially-mediated reduction of alternate electron acceptors including nitrate, sulfate and oxidized forms of iron and manganese. Following the depletion of oxygen, reduction of these various alternate electron acceptors occurs in a well-defined sequence that can affect the solubility and speciation of trace metals present in the sediment. Changes in redox status can be described in terms of both the *intensity* of reduction (Eh) and the *capacity* of sediment redox systems that buffer changes in redox intensity. Changes in metal speciation are affected by both redox intensity and capacity. Studies will be discussed that investigated changes in speciation of arsenic, selenium and chromium under different controlled Eh-pH conditions in sediment. The use of specific analytical techniques for speciation of the metals and metalloids in sediments will be presented. Differences in solubility, bioavailability and toxicity for these metals were observed following changes in redox potential. The use of the concept of "critical" redox potentials to identify important changes in speciation of these metals will also be presented. These studies demonstrate the importance of redox changes on speciation and solubility of metals and metalloids in sediment systems.

Sediments can serve as "sinks" or "sources" of toxic heavy metals to overlying water bodies. The exchange of metals across the sediment-water interface depends on a complex interplay of physical, chemical, and biological processes that control metal speciation, solubility, and the development of concentration gradients. An important feature is the gradient of metal concentration between the sediment porewater and surface water that can drive the exchange of metals across the sediment water interface. Concentrations of metals in sediment porewater, in turn, are controlled by sorption and precipitation reactions dependent on sediment properties, metal speciation, porewater composition, and fundamental physicochemical conditions such as pH and oxidation-reduction (redox) potential or Eh. Changes in sediment Eh-pH conditions can cause changes in metal speciation and solubility which can subsequently result in a flux from bottom sediments to overlying water [1].

Changes in sediment Eh-pH conditions have been shown to produce wide differences in metal and metalloid speciation and solubility under controlled conditions [2,3]. Increases in soluble metal concentrations in oxidized sediment from a salt marsh is a classic example of this process [4]. The mechanism in this particular salt marsh sediment is the oxidation of metal sulfides, including pyrite, which serve as a reservoir of toxic metals in these systems. Seasonal and temporal variation in sediment properties such as Eh and pH are not trivial occurrences in the environment. Sediments are never fully at equilibrium [5] and a wide variety of natural (e.g., tidal and non-tidal flooding) and anthropogenic (e.g., dredging) processes cause changes in these parameters. Changes in Eh conditions may not always result in predictable changes in metal speciation, however. Oxidation-reduction processes with slow kinetics are common and predictions based on an assumption of equilibrium will be incorrect. Advanced speciation models (e.g., GEOCHEM [6] and MINTEQ [7]) have been developed but these rely on equilibrium concepts and this limitation must be recognized in their use. For many metals and metalloids, there is no substitute for making accurate, direct measurements of species.

Sediment Eh status is a complex, difficult to measure parameter. While Eh can be classically defined in terms of a single chemical system (e.g., using the Nernst equation), sediment porewater is a complex mixture of compounds undergoing redox reactions in various stages of non-equilibrium. Measurements of redox potentials in these systems, therefore, represent "mixed" potentials with contributions from several redox couples [5] which may or may not be fully responsive to the Eh electrode. Measurement of Eh is commonly performed using platinum electrodes. Limitations in using platinum electrodes have been widely discussed [8], particularly as they pertain to measurement of Eh in groundwater. Despite these limitations, redox measurements are useful in sediments with high concentrations of redox-active species, especially when the electrode measurements are coupled with direct measurements of redox couples.

The Eh profile in sediments develops in response to the oxygen gradient at the sediment-water interface. Oxygen availability in sediments results from the balance of oxygen supply and demand. Sedimentation of readily-degradable organic matter generally drives O_2 consumption. When oxygen is absent, microorganisms utilize sequentially alternate electron acceptors such as nitrate, iron, manganese, sulfate and CO_2, subsequently reducing these compounds in the process. Accumulation of these reduced species in the porewater is responsible for the decreasing Eh profile in sediment. It has been shown that a critical Eh exists for the utilization of each of these alternate electron acceptors by microorganisms [9]. These transformations occur sequentially according to thermodynamic considerations.

Many of the various processes controlling the solubility and speciation of metals have been reviewed previously [10]. This paper will discuss studies documenting changes in metal speciation following changes in sediment Eh status, emphasizing the concepts of redox intensity and capacity. Obviously, other controlling variables (e.g., pH) and processes (e.g., sorption/desorption) play important roles in metal speciation and solubility. This paper will focus on cases where oxidation-reduction of metal species directly affects solubility.

2. INTENSITY AND CAPACITY OF Eh

A commonly overlooked characteristic of redox-controlled processes is that they have both *intensity* and *capacity* aspects. Measurements of redox potential with a platinum electrode measure reducing intensity but not capacity. Examples of useful intensity measurements include the "critical" redox potentials at which the inorganic redox systems become unstable (Figure 1). Although this is valuable information, it gives no indication of the size of the pool of electrons able to be accepted or donated and therefore, the amount of microbial respiration that can be supported by the redox system.

Redox capacity refers to the amount of electrons that must be added (oxidative capacity) or removed (reductive capacity) to achieve a given Eh. Redox capacity is analogous to alkalinity with respect to the

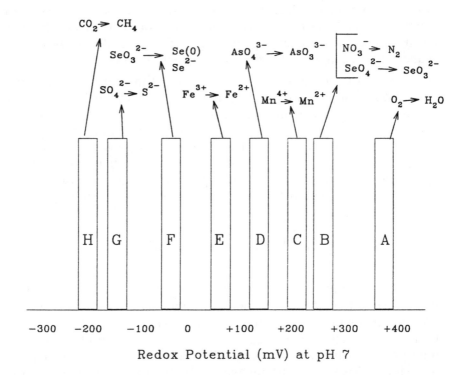

Figure 1. *Critical Redox Potentials for Transformations of Inorganic Species [2,3, 9, 11-19].*

proton condition. Methods of estimating capacity have been developed (e.g., Heron *et al.* [20]). The approach is to measure the content of the electron accepting or donating redox species (Table 1), expressing them in their oxygen equivalents. An example of this calculation is presented below.

Table 1.	Oxygen equivalents of inorganic oxidants in an alluvial soil. Calculations given in [15] based on oxidation of glucose.		
Oxidant	Concentration (μg/g)	Equivalent conversion on μg/g basis	Equivalents (as ppm O_2)
O_2	10 (as O_2)	1 ppm O_2/ppm O_2	10
NO_3^--N	43 (as N)	0.35 ppm N/ppm O_2	15.1
MnO_2	250 (as Mn^{4+})	3.43 ppm Mn/ppm O_2	858
Fe_2O_3	2500 (as Fe^{3+})	3.49 ppm Mn/ppm O_2	8725

To understand changes in metal speciation both intensity and capacity aspects of Eh must be considered. In certain instances, capacity and intensity terms can be used as qualitative, predictor variables for determining how speciation of metals will occur. Several intensity and capacity concepts relevant to metal speciation are discussed below.

1). *"Critical" redox intensity-* This can be defined as the measured Eh where a change in speciation is observed to occur in a particular soil. The critical Eh is an intensity measurement and will not indicate the extent or rate of the transformation, only that the redox intensity is favorable. Values of critical redox potentials for some of the key redox systems are given in Figure 1. In general, these potentials were determined using similar techniques (see Turner and Patrick [9]). Soil or sediment slurries were incubated in microcosms under conditions of controlled Eh and pH. The Eh was varied along the redox scale using techniques described previously (Patrick *et al.* [21]) and measurements of soluble redox elements were made at each Eh. The critical Eh was assigned as the redox potential where the reduction or oxidation was observed to occur. While the values of these critical redox potentials given in Figure 1 have been determined in the laboratory of the authors, many of these numbers and the general sequence have been determined and validated elsewhere (e.g., Sposito *et al.* [22]).

2).   *"Transformation" capacity of redox systems-* While the critical redox potential gives an indication of whether a metal species transformation is possible, a capacity term is necessary to estimate how many moles of the metal will actually be transformed by the redox process. Since every oxidation must be coupled with a suitable reduction, quantification of reductants which can participate in the reaction gives a capacity measurement of the potential species transformation. A simple example is the oxidation of Cr(III) to Cr(VI) in natural waters or sediments. Previous studies [23] have demonstrated that oxidized forms of manganese react with Cr(III) serving as an electron acceptor to convert Cr to its oxidized form, Cr(VI). The relative ability of soils and sediments to perform this transformation, therefore, can be estimated by measuring reducible Mn, usually as hydroquinone-reducible Mn. The measured Mn is an example of a capacity measurement that quantifies the ability of the soil to oxidize Cr(III). However, the intensity of reduction must still be favorable for the oxidation to occur. The theoretical, classical Nernst equation will predict a certain distribution of Cr species (e.g., Cr(III)/Cr(VI)) at a particular Eh. However, this equation assumes that species are at equilibrium and that a reversible reaction can occur. In sediments with low concentrations of reducible Mn, for example, the oxidation of Cr(III) is very slow, even when the redox intensity is favorable. Slow kinetics of transformation are common for many metals and the Nernst equation rarely has any quantitative predictive value after alterations in sediment Eh. Other metal species transformations are more complex than this Cr example. Multiple redox systems may be able to participate in the transformation and determining the transformation capacity for these species is difficult.

3). *"Buffering" capacity-* A second related capacity measurement is the ability of all redox systems to "buffer" the Eh below or above the critical Eh for a particular speciation transformation. The redox elements contributing to this capacity may have no direct effect on the metal, themselves, but serve to modify and diminish changes in Eh that may affect speciation. An example is the ferric-ferrous iron system, which in many soils and sediments is by far the most important redox system in terms of oxygen equivalents. This is demonstrated in Table 1 where redox components for a Mississippi alluvial soil are presented. The oxygen equivalent of these redox components can be calculated based on stoichiometry with glucose as an electron donor [15]. Calculations demonstate the importance of the solid redox components (e.g., Fe and Mn). In soils and sediments with appreciable iron content, changes in Eh are diminished or buffered by the electrons accepted or released by microbial metabolism using the iron system. This is particularly important in alterations such as flooding/drying or dredging where redox systems can buffer the Eh above the critical Eh of a particular transformation for a certain period of time.

This capacity term can be measured by estimating the sizes of important redox systems individually or by a non-specific method using an oxidant or reductant [20]. Several components of the capacity (e.g., O_2, NO_3^-, and SO_4^{2-}) are water-soluble and easily measured. However, since solid iron and manganese oxides represent the bulk of the redox capacity in many systems, methods of estimating oxidative capacity focus on estimating the fraction of these compounds that is potentially reducible. Various methods have been developed (for a review see Stumm and Sulzberger [24]), although lower concentrations of reducing agents are necessary to accurately quantify capacity in aquifers [20].

Other capacity measurements not directly related to oxidation-reduction are also important for metal speciation. The concept of acid-volatile sulfides, for example, is used to quantify a reactive pool of sulfide that can react with transition metals forming insoluble metal sulfides.

3.   CHROMIUM, SELENIUM AND ARSENIC

Chromium, selenium and arsenic represent a class of metals and metalloids that exist in multiple valence states in the sediment environment. Studies are described below which quantify speciation changes in response to alteration in sediment Eh. These metals and metalloids participate directly in redox reactions and can undergo oxidation/reduction between their various valence states. Changes in speciation are important because biogeochemical and toxicological characteristics of these elements change dramatically between metal forms. For example, chromium (VI) is approximately a thousand times more toxic than Cr(III). Selenate, the oxidized form of selenium, is poorly sorbed when compared to the selenite form [22]. Inorganic arsenic, e.g., arsenite (As(III)), is a suspected carcinogen [25] although certain organic arsenicals (e.g., arsenobetaine, the primary form of As in fish and shellfish) are relatively non-toxic [26]. Predicting these changes in speciation, therefore, becomes critical in assessing risk and potential effects.

Changes in metal and metalloid solubility in response to alterations in sediment Eh are not restricted to elements with multiple valence states (e.g., Cr, As and Se). Solubilities of other metals (e.g., Zn) can be controlled by sorption to redox-sensitive surfaces such as iron and manganese oxides [27]. Changes in Eh, therefore, can produce solubility changes in these elements since dissoluution of these oxides can occur under low redox conditions. Discussion presented below will focus on multi-valence elements that directly participate in oxidation-reduction reactions. The studies discussed here illustrate several important points concerning metal speciation under the influence of redox potential.

3.1   Chromium

Chromium generally exists in two valence states (Cr(VI) and Cr(III)) in the environment. Cr(VI) is the species of most concern due to its toxicity, mutagenicity, weaker interaction with particles and greater solubility. Numerous studies have investigated the transformation processes of Cr in various environments [28,29]. Again, the kinetics of these transformation processes are of interest since equilibrium can rarely be assumed. A recent study by Masscheleyn *et al.* [3] on chromium redox chemistry in a seasonally-flooded bottomland hardwood forest serves to illustrate some of these processes. The chemistry of chromium in the soil and floodwater of this system was dominated by oxidation-reduction and sorption reactions. Chromium was speciated using the s-diphenylcarbazide method [23] for Cr(VI) and ICP for total Cr. Trivalent Cr was determined by difference. A critical Eh of +300 mV was identified for the reduction of Cr(VI) to Cr(III) in soils. Kinetics of Cr(VI) to Cr(III) reduction were rapid (<1 minute) and complete. A high capacity for reduction was observed where 50 mg of Cr(VI)/kg of soil was instantaneously reduced. The reverse reaction, the oxidation of Cr(III) to Cr(VI), was not observed in soil but was observed in the floodwater. A surface film containing Fe and Mn was prominent in the wetland floodwater during periods of the year. Oxidation of Cr(III) has been shown to be dependent on the presence of amorphous manganese oxides [23]. The presence of these compounds in the surface film was attributed as the cause of Cr oxidation in floodwater.

The effect of changes in Cr speciation was most evident in sorption studies. Sorption is often mathematically described using isotherm equations. For Cr, however, isotherm parameters lump not only sorption but oxidation-reduction and precipitation reactions together. Therefore, what is referred to as Cr "sorption" isotherms and coefficients, actually include multiple processes. In sorption isotherms conducted under controlled Eh conditions sorption was weak (as indicated by low sorption coefficients, Freundlich "K") above the critical Eh for Cr(VI) reduction. Below the critical Eh, when Cr(III) was present, "sorption" coefficients increased by three-orders of magnitude (Figure 2).

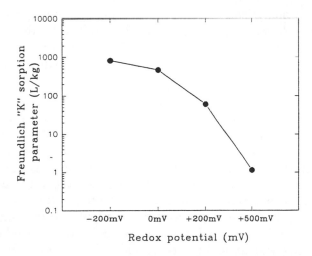

Figure 2.   *Freundlich Isotherm Parameters for Cr(VI) in Spring Bayou Bottomland Hardwood Wetland Soil Under Controlled Redox Potentials.   Data adapted from Masscheleyn et al. [3].*

   Chromium redox chemistry serves to illustrate several important points about changes in speciation with Eh. First, speciation changes are not always reversible (or have much different kinetics). Because of this, changes in bulk sediment Eh are not always accompanied by changes in metal speciation. Secondly, for Cr, relative rates of transformation processes, i.e., oxidation of Cr(III) and reduction of Cr(VI), indicate whether this environment serves as a "sink" or "source" of Cr. Since the reduction of Cr(VI) is rapid and occurs in both floodwater and soils (a large pool of Cr) and the oxidation of Cr(III) is slow and occurs only in floodwater (a small pool of Cr), this wetland has a great potential to serve as a net sink for Cr.

3.2   Selenium

   Selenium can exist in four different valence states, selenide (Se(-II)), elemental Se (Se(0)), selenite (Se(IV)), and selenate (Se(VI)). Selenium chemistry is comparable to another Group VIA element, sulfur. Again, major factors affecting the fate and transport of selenium in sediment-water systems are its oxidation-reduction and the subsequent effects on retention processes (e.g., sorption). Numerous studies have documented aspects of selenium biogeochemistry [30,31]. Studies completed by Masscheleyn *et al.* [17] on contaminated Kesterson reservoir sediment document some important features of Se geochemistry and the effect of Eh. A critical step in these studies is the ability to analytically separate different selenium species. A hydride generation/trapping/detection apparatus utilized for this study has been described in detail elsewhere [32]. It distinguishes between selenate (Se(VI)), selenide (Se(IV)), dimethyl selenide, oxidized methylated Se compounds, and reduced (Se(0,-II)) selenium.

   In Kesterson sediment, solubility of Se under reduced conditions was lower (controlled by an iron selenide phase) than under oxidizing conditions. When Eh is increased, oxidation of Se(II,0) to selenite (Se(IV)) was rapid (time scale of several days). Subsequent oxidation to the more oxidized selenate (Se(VI)) species was much slower but given sufficient time, under oxidizing conditions (i.e., several weeks), selenate is the dominant species in solution. Reversibility of these transformations was not considered explicitly in this study, however. Critical Ehs for the transformations were determined and were +50 mV for the Se(-II,0) oxidation to selenide, Se(IV) and +200 mV for the oxidation of selenide to selenate, Se(VI). Additional studies using Hyco Reservoir, NC sediments [2] confirmed these results. In this sediment, when Eh was raised to +500 mV, 68% of the total Se present was solubilized. Biomethylation of Se species was also important under oxidized and moderately reduced conditions (500, 200 and 0 mV).

Selenium redox chemistry demonstrates several features about changes in metal speciation in response to changes in Eh. Again, differences in the kinetics of transformation between different species affected the predictability of speciation. Release of high percentages (e.g., > 50%) of total selenium was observed when sediment was oxidized. Conversion of selenium to organic forms was also observed. Dramatic differences in the solubility of oxidized species such as the selenium oxyanions and the reduced species (metal selenides) emphasizes the importance of considering speciation changes in the geochemistry of this metalloid.

3.3 Arsenic

Arsenic can exist in two primary forms in soils, arsenate (As(V)) and arsenite (As(III)). Both forms are also subject to biomethylation, resulting in organic arsenic species. Oxidation-reduction, and its subsequent effects on sorption/precipitation are major factors affecting the fate of arsenic in sediment-water systems. Studies by Masscheleyn [2,14] document some important features of arsenic geochemistry in soils and sediments. Arsenic was speciated using a similar hydride generation technique as used for selenium [33]. The method measures nanogram quantities of arsenite (As(III)), arsenate (As(V)), monomethylarsonic acid (MMAA) and dimethylarsinic acid (DMAA).

In contrast to selenium, arsenic solubility was shown to be greater under anaerobic conditions than under aerobic conditions in a contaminated Kolin, Louisiana soil. Up to 40% of the total arsenic present in the soil became mobilized after the change in Eh to reducing conditions. This was attributed primarily to release of As(V) following dissolution of iron oxyhydroxides under reducing conditions. As(V) had the highest species concentrations under oxidizing conditions while As(III) had the highest species concentration under reducing conditions. Although not thermodynamically favorable, significant concentrations of As(V) were observed under reducing conditions after the observed dissolution of iron oxyhydroxides in the contaminated Kolin, Louisiana soil. As(V) was detectable even after five weeks and after flooding and subsequent changes in Eh. Slow kinetics of the As(V) to As(III) reduction coupled with the release of high concentrations of Mn upon reduction made the precipitation of a $Mn_3(AsO_4)_2$ phase under reduced conditions a likely event as indicated by GEOCHEM.

In a contaminated Hyco Reservoir, NC sediment the same general trend was observed [2]. Approximately 51% of the total As was remobilized under anaerobic conditions. In this sediment, however, porewater was undersaturated with respect to As minerals (e.g., $Mn_3(AsO_4)_2$). Other factors, (e.g., sorption) likely controlled As solubilities under reducing conditions.

Arsenic redox chemistry demonstrates several features about changes in metal speciation in response to changes in Eh. Slow kinetics of metal transformations were again observed (As(V) to As(III) reduction). The strong dependence of arsenic speciation on the chemistry of the iron and manganese system were demonstrated. In these soils and sediments, the pools of iron and manganese accounted for the bulk of the transformation and buffering capacity of the system. Speciation changes in As were according to thermodynamic expectations, but only qualitatively. Significant deviation from the Nernst equation was observed even after weeks of incubation and predictions were qualitative at best. Finally, As is an example of a metalloid whose solubility is higher under reducing conditions. This is in contrast to the bulk of transition metals whose solubility is dramatically lower under reducing conditions as metal sulfides are formed.

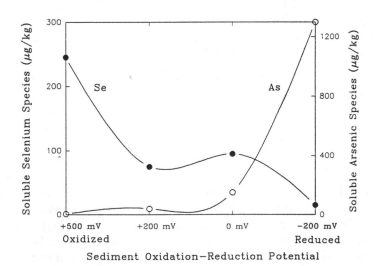

Figure 3.   *Soluble Arsenic and Selenium Species in a Hyco Reservoir, NC Sediment Under Controlled Redox Conditions at Natural pH (4.0 for 500 mV, 5.3 for 200 mV, 6.1 for 0 mV, and 6.9 for -200 mV). Data adapted from Masscheleyn et al. [2].*

4.   CONCLUSIONS

Changes in sediment Eh in response to natural or anthropogenic alteration is an important control on metal solubility. Unfortunately, the variety of conditions and elements resist generalizations about the effect of Eh on metal speciation and solubility. It has been stated that oxidizing conditions favor metal insolubility while reducing conditions favor metal solubility or mobility [10]. There are at least several important exceptions to this rule (e.g., Cr and Se as discussed in this paper) and probably many others. In coastal sediments or other areas with high concentrations of sulfides, the opposite trend (greater solubility under aerobic conditions) is observed for many metals. The studies discussed above demonstrate other complicating features. First of all, the Nernst equation, the classical redox equilibrium equation, has only limited predictive ability for metals whose transformation kinetics are slow. In addition, the dynamic nature of many soils and sediments makes an equilibrium approach difficult to justify. In these cases speciation changes must be modeled kinetically, thus requiring the determination of a transformation rate. Secondly, many trace metal transformations require specific oxidants or reductants that may not be present in a given soil or sediment. This leads to irreversibility or very slow kinetics for certain metal transformations, making the Nernst equation, where equilibrium is assumed, completely unusable.

Predicting changes in metal speciation in response to Eh, then requires understanding the transformation, itself. What is the critical Eh intensity for the oxidation-reduction transformation? What is the capacity of the oxidant or reductant(s) in the sediment which actually drive the transformation? The buffering capacity, or how fast bulk Eh changes occur in response to a change in conditions, must also be understood. Predicting these transformations will require better techniques of measuring Eh intensity and capacity. Monitoring certain easily-measured redox couples (e.g., Fe^{+2}/Fe^{+3}, Mn^{+2}/Mn^{+4}) may indicate the possibility of a metal speciation transformation in a sediment, particularly if these indicator couples are directly involved in the reaction. This will be more accurate than using bulk Eh using platinum electrodes which are limited by non-equilibrium of many redox-couples in the system.

Finally, it should be recognized that any rate depends on a large number of variables, of which Eh is only one. Generalizing across groups of metals and environments will be difficult because of the variables contributing to the kinetic rate term. Better methods of measuring some of these variables (e.g., Eh) may improve predictability and improve our understanding of metal flux from sediments undergoing changes in Eh.

REFERENCES

1. Kerner, M. and K. Wallman, "Remobilization Events Involving Cd and Zn from Intertidal Mud Flat Sediments in the Elbe Estuary During the Tidal Cycle," *Est. Coast. Shelf Sci.* 35:371-393 (1992).

2. Masscheleyn, P.H., R.D. DeLaune and W.H. Patrick, Jr., "Arsenic and Selenium Chemistry as Affected by Sediment Redox Potential and pH," *J. Environ. Qual.* 20:522-527 (1991).

3. Masscheleyn, P.H., J.H. Pardue, R.D. DeLaune and W.H. Patrick, Jr., "Chromium Redox Chemistry in a Lower Mississippi Valley Bottomland Hardwood Wetland," *Environ. Sci. Technol.* 26:1217-1226 (1992).

4. DeLaune, R.D. and C.J. Smith, "Release of Nutrients and Metals Following Oxidation of Freshwater and Saline Sediment," *J. Environ. Qual.* 14:164-168 (1985).

5. Stumm, W. and J.J. Morgan, *Aquatic Chemistry 2nd edition.* John Wiley & Sons, (New York, 1981).

6. Sposito, G. and S.V. Mattigod, "GEOCHEM: A Computer Program for the Calculation of Chemical Equilibria," in *Soil Solutions and Other Natural Waters*, University of California, (Riverside, CA 1979).

7. Brown, D.S. and J.D. Allison, *MINTEQA1 Equilibrium Metal Speciation Model: A User's Manual*, EPA/600/3-87/012, U.S. EPA, (Athens, GA 1987).

8. Lindberg, R.D. and D.D. Runnells, "Ground Water Redox Reactions: An Analysis of Equilibrium State Applied to Eh Measurements and Geochemical Modeling," *Science* 225:925-927 (1984).

9. Turner, F.T. and W.H. Patrick, Jr., "Chemical Changes in Waterlogged Soils as a Result of Oxygen Depletion," in *Trans. 9th Int. Cong. Soil Sci.*, *Adelaide*, J.W. Holmes Ed., pp. 53-65. Elsevier, (New York, 1968).

10. McLean, J.E. and B. Bledsoe, "Behavior of Metals in Soils," *Ground Water Issue. Environmental Protection Agency*, p. 24 RSKERL, (Ada, OK 1992).

11. Reddy, K.R. and W.H. Patrick, Jr., "Effect of Alternate Aerobic and Anaerobic Conditions on Redox Potential, Organic Matter Decomposition, and Nitrogen Loss in a Flooded Soil," *Soil Biol. Biochem.* 7:87-94 (1975).

12. Patrick, W.H., Jr., "Nitrate Reduction Rates in a Submerged Soil as Affected by Redox Potential," *7th Int. Cong. Soil Sci., Madison, WI* 2:494-500 (1960).

13. Buresh, R.J. and W.H. Patrick, Jr., "Nitrate Reduction to Ammonium and Organic Nitrogen in an Estuarine Sediment," *Soil Biol. Biochem.* 13:279-283 (1981).

14. Masscheleyn, P.H., R.D. DeLaune and W.H. Patrick, Jr., "Effect of Redox Potential and pH on Arsenic Speciation and Solubility in a Contaminated Soil," *Environ. Sci. Technol.* 25:1414-1419 (1991).

15. Patrick, W.H., Jr., "The Role of Inorganic Redox Systems in Controlling Reduction in Paddy Soils," in *Proc. Symp. on Paddy Soils*, Inst. of Soil Sci. Acad., Sinica, China Ed., pp. 107-117 Springer-Verlag, (New York, 1981).

16. Patrick, W.H., Jr., "Extractable Iron and Phosphorus in a Submerged Soil at Controlled Redox Potentials," *8th Int. Cong. Soil Sci., Bucharest, Romania* IV:605-608 (1964).

17. Masscheleyn, P.H., R.D. DeLaune and W.H. Patrick, Jr., "Transformations of Selenium as Affected by Sediment Oxidation-Reduction Potential and pH," *Environ. Sci. Technol.* 24:91-96 (1990).

18. Connell, W.E. and W.H. Patrick, Jr., "Sulfate Reduction in Soil: Effects of Redox Potential and pH," *Science* 159:86-87 (1968).

19. Masscheleyn, P.H., R.D. DeLaune and W.H. Patrick, Jr., "Methane and Nitrous Oxide Emissions from Laboratory Measurements of Rice and Soil Suspensions: Effect of Soil Oxidation-Reduction Status," *Chemosphere* 26:251-260 (1992).

20. Heron, G., T.H. Christensen and J.C. Tjell, "Oxidative Capacity of Aquifer Sediments," *Environ. Sci. Technol.* 28:153-158 (1994).

21.   Patrick, W.H., Jr., B.G. Williams and J.T. Moraghan, "A Simple System for Controlling Redox Potential and pH in Soil Suspensions," *Soil Sci. Soc. Am. J.* 36:573-576 (1973).

22.   Sposito, G., A. Yang, R.H. Neal and A. Mackzum, "Selenate Reduction in an Alluvial Soil," *Soil Sci. Soc. Am. J.* 55:1597-1602 (1991).

23.   Bartlett, R. and B. James, "Behavior of Chromium in Soils: III. Oxidation," *J. Environ. Qual.* 8:31-35 (1979).

24.   Stumm, W. and B. Sulzberger, "The Cycling of Iron in Natural Environments: Considerations Based on Laboratory Studies of Heterogeneous Redox Processes," *Geochim. Cosmochim. Acta* 56:3233-3257 (1992).

25.   WHO, *Environmental Health Criteria 18, Arsenic.* World Health Organization, (Geneva 1981).

26.   FAO/WHO, "Toxicological Evaluation of Certain Additives and Contaminants," *The 33rd Meeting of the Joint FAO/WHO Expert Committee on Food Additives*, (Geneva 1989).

27.   Gambrell, R.P. and W.H. Patrick, "Cu, Zn And Cd Availability in a Sludge-amended Soil Under Controlled pH and Redox Potential Conditions," in *Inorganic Contaminants in the Vadose Zone, Ecological Studies 74*, B. Bar-Yosef, N.J. Barrow and Y. Goldshmidt, Eds., Springer-Verlag, (Berlin, 1989).

28.   Anderson, L.D., D.B. Kent and J.A. Davis, "Batch Experiments Characterizing the Reduction of Cr(VI) Using Suboxic Material from a Mildly Reducing Sand and Gravel Aquifer," *Environ. Sci. Technol.* 28:178-185 (1994).

29.   Bartlett, R. and J.M. Kimble, "Behavior of Chromium in Soils: II. Hexavalent Forms," *J. Environ. Qual.* 5:383-386 (1976).

30.   Oremland, R.S., J.T. Hollibaugh, A.S. Maest, T.S. Presser, L.G. Miller and C.W. Culbertson, "Selenate Reduction to Elemental Selenium by Anaerobic Bacteria in Sediments and Culture: Biogeochemical Significance of a Novel, Sulfate-Independent Respiration," *Appl. Environ. Microbiol.* 55:2333-2343 (1989).

31.   Neal, R.H., G. Sposito, K.M. Holtzclaw and S.J. Traina, "Selenite Adsorption on Alluvial Soils: I. Soil Composition and pH Effects," *Soil Sci. Soc. Am. J.* 51:1161-1165 (1987).

32. Masscheleyn, P.H., R.D. DeLaune and W.H. Patrick, Jr., "Selenium Speciation in Aqueous Solutions Using a Hydride Generation Atomic Absorption Spectrophotometry Technique," *Spectroscopy Lett.* 24:307-322 (1991).

33. Masscheleyn, P.H., R.D. DeLaune and W.H. Patrick, Jr., "A Hydride Generation Atomic Absorption Technique for Arsenic Speciation," *J. Environ. Qual.* 20:96-100 (1991).

8

DYNAMICS OF TRACE METAL INTERACTIONS WITH AUTHIGENIC SULFIDE MINERALS IN ANOXIC SEDIMENTS

John W. Morse
Department of Oceanography
Texas A&M University
College Station, TX 77843

1.   INTRODUCTION

There is growing observational and experimental evidence that interactions between toxic metals and authigenic sulfide minerals in anoxic sediments may play a major role in controlling the bioavailability of these metals. The coprecipitation and/or adsorption of many toxic trace metals with pyrite and acid volatile sulfide minerals (AVS) may effectively sequester the metals from uptake by organisms. However, during seasonal migration of the redoxcline, and resuspension and dredging of sediments, sulfide minerals can be oxidized and release associated toxic metals. Consequently, the role of sedimentary sulfides in influencing the bioavailability of toxic metals is complex and some current approaches to this problem are probably overly simplistic.

Sediments are the dominant sink for trace metals of environmental concern in most aquatic environments. It is consequently important to understand the chemical behavior of trace metals in sediments in order to assess their potential impact on benthic biota or when sediments are resuspended in the water column. The fate of trace metals in sediments is controlled by a complex interplay of biological, chemical, and physical factors that leads to extensive transformation of their speciation under the often rapidly changing pH and redox conditions encountered below the sediment-water interface.

Bacterial reduction of sulfate to oxidize organic matter is frequently a major process leading to the production of H_2S in anoxic sediments. Many transition and heavy metals form highly insoluble sulfide minerals, and have been observed to adsorb and coprecipitate with commonly formed authigenic iron sulfide minerals. Consequently, considerable interest has been generated in the possibility that the interaction of toxic trace metals with the sedimentary sulfide system may substantial reduce their availability to organisms. Much of the current interest in this subject is centered around the observation that there is no toxicity of cadmium to amphipods when the molar concentration of cadmium is less than that of reduced sulfur [1].

My students and I have been studying the interaction of metals with sedimentary iron sulfide minerals for close to a decade. Our research has taken us down several paths; both in experimental laboratory studies and in field studies, primarily in marine and estuarine environments. This paper will present a brief general review of the chemistry of sulfidic sediments, and a synthesis of the work that our group has produced.

2.   OVERVIEW OF SULFIDE CHEMISTRY IN ANOXIC SEDIMENTS

Sediments, in which sulfate reduction results in the production of sulfide minerals relatively close to the sediment-water interface, typically, but not exclusively, contain a major fraction of silt and clay sized terrigenous material, greater than 0.5 wt.% organic-C and accumulate at moderate to high sedimentation rates. Such sediments are common to lakes, estuaries, and marine shelf and slope environments. Since the early 1970's the major processes resulting in formation of sedimentary sulfide minerals have been known [2-4]. These are:

1)   Bacterially-mediated oxidation of organic matter via sulfate reduction leading to production of H_2S.

organic matter $+ SO_4^{2-}$ -bacteria $\rightarrow$ H_2S

2)   Formation of metastable sulfide minerals via interactions between H_2S and sedimentary iron minerals that are capable of relatively rapidly reacting with H_2S. These metastable iron minerals may include amorphous-FeS, mackinawite (FeS), and greigite (Fe_3S_4). Upon HCl leaching of sediment these minerals are

dissolved and evolve H_2S. Therefore, they are operationally grouped as acid volatile sulfides (AVS).

$$\text{reactive-Fe} + H_2S \rightarrow AVS$$

3) The dominant form of reduced sulfide minerals in most sediments is thermodynamically stable pyrite (FeS_2). In order for pyrite to form, some of the H_2S must be partially oxidized by bacteria to elemental sulfur.

$$H_2S + O_2 \text{ -bacteria} \rightarrow S^o$$

4) The AVS minerals can then react with S^o via complex pathways to form pyrite.

$$AVS + S^o \rightarrow \text{ pyrite}$$

During the past two decades hundreds of papers have been written further elaborating the details of these processes. This has led to a growing appreciation of the complexity of the sedimentary sulfide system [5,6]. Although the basic conceptual framework has remained largely intact, numerous questions have arisen (e.g., can pyrite form directly in sediments from the reaction of dissolved Fe^{2+} with polysulfide ions?).

Interest in the interaction of toxic trace metals with the sedimentary sulfide system can largely be justified on the basis of two fundamental considerations. The first is that processes leading to formation of sedimentary sulfide minerals are also closely associated with the destruction of organic and inorganic components of sediments with which potentially bioavailable fractions of toxic metals are strongly associated. This frees these metals to adsorb or precipitate within the sediment, or be transported to overlying waters. The second is that most metals of interest can form highly insoluble sulfide minerals and have been observed to extensively coprecipitate with iron sulfide minerals. Consequently, the bioavailability of toxic trace metals in anoxic sediments may be largely controlled by the formation of either discrete toxic metal sulfide minerals or their coprecipitation with or adsorption on other sulfide minerals.

3.   TRACE METAL INTERACTIONS WITH PYRITE

3.1   Pyritization of metals

The interaction of trace metals with pyrite will be considered first. This is because in most sulfidic sediments pyrite is the thermodynamically stable and by far most abundant sulfide mineral. Also, considerably more progress has been made in understanding the interactions of metals with pyrite than with other iron sulfide minerals.

Early attempts to quantify the concentration of trace metals in sedimentary pyrite relied primarily on two techniques. The first was SEM/EDAX and electron microprobe analysis which was restricted to relatively large individual pyrite grains and was very often not sensitive enough to determine the concentration of metals of interest [7-10]. The second was heavy liquid separation. Heavy liquid separation techniques did not yield quantitative recovery of pyrite (generally only 2% to 25%) and other heavy minerals were recovered with pyrite [11-13]. Additionally, major questions persisted as to how well such analyses could be extrapolated to total metal concentrations associated with pyrite in sediments.

In 1982, Lord [14] developed a sequential extraction technique for the determination pyrite-Fe. This technique consists of initial leaches of the sediment with citrate-dithionite (generally considered the "reactive" fraction) and HF (the silicate fraction), and final extraction of pyrite with nitric acid. We [15] spent considerable time and effort refining this technique further for determination of pyrite associated trace metals. Major modifications included using HCl instead of citrate-dithionite as the initial leachate, because of contamination and analytical problems associated with citrate-dithionite for some metals, and adding an NaOH digestion, to further remove organic matter, prior to the nitric acid digestion of pyrite.

This technique was initially applied to the study of diagenesis of metals with depth in sediments. Several of the metals studied generally exhibited clear trends of substantial transfer from the HCl-leachable fraction (subsequently referred to as the reactive fraction) to the pyrite fraction near the sediment-water interface. Examples from Baffin Bay, Texas, are presented in Figure 1 [16]. Some metals (e.g., As and Hg) exhibited almost total pyritization near the sediment-water interface, whereas other metals (e.g., Zn) exhibited only a limited tendency to become associated with the pyrite fraction.

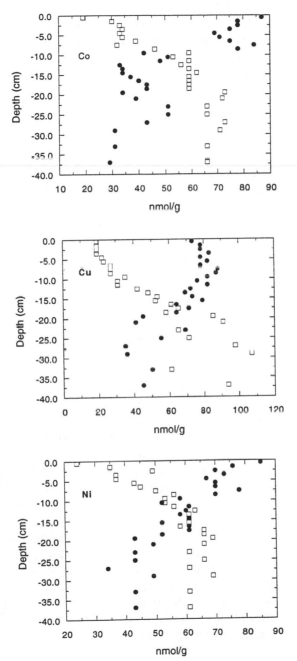

Figure 1.   **Examples of Diagenetic Transformation of Metals with Depth in Anoxic Sediments from the Reactive Fraction (1N HCl leachable, open squares) to the Pyrite Fraction (solid circles).**

The extent to which sedimentary Fe phases, that have a potential to supply Fe for pyrite formation, are transformed to pyrite during diagenesis has been defined as the degree of pyritization (DOP) [3]. DOP can be expressed as

$$DOP = \frac{Pyrite\text{–}Fe}{Pyrite\text{–}Fe + Reactive\text{–}Fe} \tag{1}$$

This concept was expanded to metals (Me) other than Fe defining the degree of trace metal pyritization (DTMP) similarly to DOP for Fe [16]. By comparing DTMP with DOP it is possible to relate the pyritization of a given trace metal to that of Fe, which is the dominant metal in pyrite.

$$DTMP = \frac{Pyrite\text{–}Me}{Pyrite\text{–}Me + Reactive\text{–}Me} \tag{2}$$

The general relationships between the DTMP of trace metals and DOP, based on several thousand analyses [17], are shown in Figure 2. Although occasional exceptions to these general trends have been observed, the general pattern of trace metals behavior is remarkably constant in a wide variety of aquatic environments. The tendency of transition metals to closely follow the behavior of Fe is not surprising since Fe is also a transition metal. However, chemical pathways may be influenced by the differing redox behavior of these metals (e.g., Fe goes from +3 to +2, whereas Cu probably goes from +2 to +1). The strong tendency of As and Hg to become associated with the pyrite fraction is likely to be the result of quite different processes. As is likely to be reduced from arsenate to arsenite and incorporated as an anion in analogous fashion to formation of arsenopyrite. Hg forms one of the most insoluble sulfide minerals (cinnabar) known and is likely to have an extremely large partition coefficient in pyrite and a high affinity for the surface of pyrite. The general low extent of association of class B metals with the pyrite fraction is perplexing. Possible reasons for this behavior include complexing of the metals in solution, coprecipitation or adsorption with other phases, such as calcium carbonate, and formation of authigenic carbonate, sulfide and phosphate minerals of these metals.

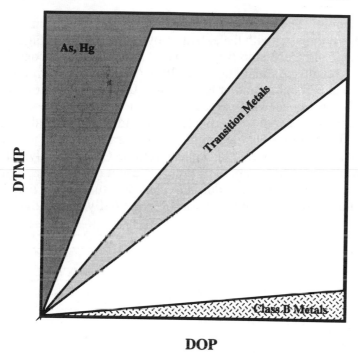

DOP

Figure 2.   *General Relationship of DTMP to DOP for Different Metals.*

3.2   Oxidative release of pyritized metals

Experimental studies [18] have demonstrated that all sedimentary metastable iron sulfide minerals and a surprisingly large (typically ~20% or greater) fraction of sedimentary pyrite can be rapidly oxidized when anoxic sediments are exposed to oxic seawater. This raises the possibility that during sediment resuspension or seasonal migration of the redoxcline, oxidation of iron sulfide minerals may result in the release of toxic metals from a previously unrecognized pool of potentially bioavailable metals.

We are still in the early stages [19] of our research on the oxidative release of trace metals associated with the pyrite fraction of sediments when sediments are exposed to oxic seawater. Sediments recovered from near (within ~10 cm) the sediment-water interface have been used in experimental studies. Initial observations indicate that, for the three metals studied (As, Cu, and Hg), usually at least a 25% oxidation of the pyritized metal occurred during one day's exposure to oxic seawater. However, the extent of metal oxidation was highly variable for both a given metal, between different samples and for different metals within a single sample. The extent of a loss of As, Cu, and Hg from the pyrite fraction is compared with the extent of pyrite oxidation in Figure 3, for sediments from Galveston Bay, Texas. Although there is considerable scatter in results, As and Hg generally exhibit a loss from the pyrite fraction that is roughly similar to the extent of pyrite oxidation, whereas Cu loss is often over twice as great. As and Hg released from oxidation of pyrite in one day are approximately one and a half times that of their reactive fraction (1N HCl leachable) concentrations and Cu release is about twice its reactive fraction concentration. Consequently, although these results must be regarded as preliminary, it appears that oxidative release of metals from pyrite during sediment resuspension may have a major impact on the concentration of potentially bioavailable toxic metals.

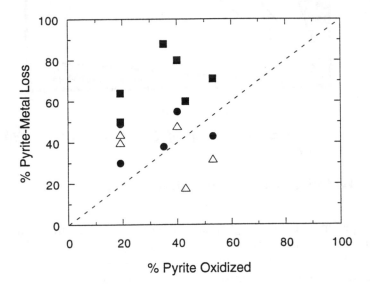

Figure 3.   *Comparison of Cu (solid circles), As (solid squares), and Hg (open triangles) Release from the Pyrite Fraction of Anoxic Sediments Exposed to Oxic Seawater for One Day to the Percentage of Sedimentary Pyrite Oxidized.*

4.   TRACE METAL INTERACTIONS WITH MACKINAWITE

As previously discussed, amorphous-FeS, mackinawite and greigite are the metastable sulfide minerals believed to comprise AVS. Amorphous-FeS is, based on experimental studies [20], likely to be relatively short-lived in sediments where it should transform to mackinawite within hours to days. Greigite, although common in organic-rich salt marsh sediments, has rarely been identified in marine and estuarine sediments [21]. Consequently, the study of the interaction of metals with mackinawite may provide a good approach to understanding their interactions with AVS in anoxic sediments.

Natural mackinawite is commonly found in association with sulfide ores, where it exhibits considerable concentrations of other transition metals and can take up significant concentrations of cations with similar atomic radii to iron. Although it has been possible to develop techniques for close to quantitative extraction of metals associated with authigenic sedimentary pyrite [15], there are currently no reliable methods for obtaining similar data for sedimentary mackinawite. The primary reasons for this are that mackinawite is easily oxidized during sample handling, it is generally very finely distributed, often as coatings on other minerals, and it is relatively easily dissolved by reagents commonly used to obtain various "reactive" metal phases [22].

Observations indicating mackinawite can accommodate coprecipitated cations, current inability to directly determine the composition of sedimentary mackinawite, and its common occurrence in anoxic sediments, led us to conduct experimental measurements of metal sorption and coprecipitation reactions with mackinawite [23]. A major difficulty encountered in these studies was that mackinawite is generally several orders of magnitude more soluble than most other metal sulfide phases of interest. Therefore, solutions supersaturated or in equilibrium with mackinawite are usually highly supersaturated with respect to other metal sulfide phases (except MnS which is more soluble than mackinawite). The possibility that pseudo-homogeneous nucleation of the metal sulfide minerals from supersaturated solutions would not occur over the time necessary to perform the experiments was consequently investigated. Manganese, cobalt and nickel were the only metals that behaved in an acceptable manner. Although other metals generally did not rapidly precipitate, their removal rate from solution was too rapid to perform our experiments. However, the addition of a small amount of mackinawite to solutions resulted in rapid metal removal to concentrations below analytical detection limits. This indicates that the metals were probably rapidly adsorbed on mackinawite.

We found that Langmuir-type plots could be used to determine the apparent adsorption constant, which decreases with increasing surface affinity. Apparent adsorption constants decrease in the order of $Mg^{2+} \approx Ca^{2+} \approx Mn^{2+} > Co^{2+} > Ni^{2+}$. Divalent metals whose sulfides are more soluble than mackinawite, thus have similar surface affinities, whereas those whose sulfide phases are less soluble than mackinawite have increasing surface affinities that correlate in order of metal sulfide solubility.

Partition coefficients were also determined for Mn, Co, and Ni in mackinawite using a chemostat technique. The partition coefficients increase with decreasing solubilities of these metal sulfides as expected. However, in the cases of Co and Ni there are large deviations from the ideal thermodynamic distribution coefficient. A possible explanation is that highly non-ideal solid solutions are formed. If the trend of increasing partition coefficients with decreasing metal sulfide mineral solubilities, observed for Mn^{2+}, Co^{2+}, and Ni^{2+}, is even grossly followed, then extremely large partition coefficients for many metals in mackinawite are probable.

5.   SUMMARY AND DISCUSSION

Observed associations of many trace metals with the pyrite fraction of anoxic sediments, as well as toxicological studies and experimental results on the relationship between AVS minerals and toxic metals, indicate that the sedimentary sulfide system probably plays a major role in controlling the potential bioavailability of many toxic metals. Although the role of sedimentary sulfide minerals in influencing the potential bioavailability of toxic metals is primarily viewed as removing them from biologically available chemical species, some care should be used in generalizing this influence. This is because of the highly dynamic nature of the sedimentary sulfide system near the sediment-water interface.

Bioturbation and bioirrigation of anoxic sediments, as well as reactions between H_2S and oxides, results in over 90% of the produced reduced sulfides in most sediments being oxidized [24]. Consequently, metals that become associated with sulfide minerals may be subsequently released via oxidation within the sediments. Additionally, the depth of the redoxcline can undergo substantial vertical migration within sediments on a seasonal basis. The concentrations of pyrite and AVS may consequently undergo large variations at a given depth below the sediment-water interface [25]. Other processes that lead to sediment resuspension, such as storms, bottom trawling, and dredging activities can result in a rapid and complete oxidation of associated AVS, and a substantial portion of sedimentary pyrite may also be oxidized.

These processes lead to a complex and dynamic environment for metals near the sediment-water interface. In this environment, toxic metals may undergo repeated transformations between oxic and sulfidic phases. It is, therefore, over simplistic to consider that an association of toxic metals with AVS or pyrite removes them from potential bioavailability.

REFERENCES

1.   Di Toro, D.M., J.D., Mahony, D.J. Hansen, K.J. Scott, M.B. Hicks, S.M. Mayr and M.S. Redmond, "Toxicity of Cadmium in Sediments: The Role of Acid Volatile Sulfide," *Environ.Toxicol. Chem.* 9:1487-1502 (1990).

2.   Rickard, D.T., "The Chemistry of Iron Sulfide Formation at Low Temperatures," *Stockholm Contrib. Geol.* 20:67-95 (1969).

3.   Berner, R.A., "Sedimentary Pyrite Formation," *Amer. J. Sci.* 268:1-23 (1970).

4.   Sweeney, R.E. and I.R. Kaplan, "Pyrite Framboid Formation: Laboratory Synthesis And Mad Marine Sediments," *Geol.* 68:618-634 (1973).

5.   Morse, J.W., F.J. Millero, J.C. Cornwell and D.T. Rickard, "The Chemistry of the Hydrogen Sulfide and Iron Sulfide Systems in Natural Waters," *Earth-Sci. Rev.* 24:1-42 (1987).

6.   Luther, G.W. III and T.M. Church, "An Overview of the Environmental Chemistry of Sulphur in Wetland Systems," in *SCOPE 48-Sulphur Cycling on the Continents: Wetlands, Terrestrial Ecosystems amd Associated Water Bodies*, R.W. Howarth, J.W.B. Stewart and M.V. Ivanov, Eds., pp. 125-144 John Wiley & Sons, (New York, 1992).

7.   Luther, G.W. III, A.L. Meyerson , J.J. Krajewski and R. Hires, "Metal Sulfides in Estuarine Sediments," *J. Sediment. Petrol.* 52:664-666 (1980).

8.   Lee, F.Y. and J.A. Kittrick, "Electron Microprobe Analysis of Elements Associated with Zinc and Copper in an Oxidizing and An Anaerobic Soil Environment," *Soil Sci. Am. J.* 48:548-554 (1984).

9.   Jacobs, L., S. Emerson and J. Skei, "Partitioning and Transport of Metals Across the O_2/H_2S Interface in a Permanently Anoxic Basin: Framvaren Fjord, Norway," *Geochim. Cosmochim. Acta* 49:1433-1444 (1985).

10. Cabri, L.J., J.L. Campbell, J.H.G. Laflame, R.G. Leigh, J.A. Maxwell and J.D. Scott, "Proton-microprobe Analysis of Trace Elements in Sulfides from Some Massive-sulfide Deposits," *Can. Mineral.* 23:133-148 (1985).

11. Raiswell, R. and J. Plant, "The Incorporation of Trace Elements into Pyrite During Diagenesis of Black Shales, Yorkshire, England," *Econ. Geol.* 75:684-689 (1980).

12. Roberts, F.I., "Trace Element Chemistry of Pyrite: A Useful Guide to the Occurrence of Sulfide Base Metal Mineralization," *J. Geochem. Explorat.* 17:49-62 (1982).

13. Volkov, I.I. and L.S. Fomina, "Influence of Organic Material and Processes of Sulfide Formation on Distribution of Some Trace Elements in Deep-water Sediments of Black Sea," in *The Black Sea, Geology, Geochemistry and Biology*, E.T. Degens and D.A. Ross, Eds., pp. 456-477 Mem. 20. American Association of Petroleum Geologists (1974).

14. Lord, C.J., "A Selective and Precise Method for Pyrite Determination in Sedimentary Materials," *J. Sediment. Petrol.* 52:664-666 (1982).

15. Huerta-Diaz, M.A. and J.W. Morse, "A Quantitative Method for Determination of Trace Metal Concentration in Sedimentary Pyrite," *Marine Chem.* 29:119-144 (1990).

16. Morse, J.W., B.J. Presley, R.J. Taylor, G. Benoit and P. Santschi, "Trace Metal Chemistry of Galveston Bay: Water, Sediments and Biota," *Mar. Envir. Res.* 36:1-37 (1993).

17. Huerta-Diaz, M.A. and J.W. Morse, "The Pyritization of Trace Metals in Anoxic Marine Sediments," *Geochim. Cosmochim. Acta* 56:2681-2702 (1992).

18. Morse, J.W., "Sedimentary Pyrite Oxidation Kinetics in Seawater," *Geochim. Cosmochim. Acta* 55:3665-3668 (1991).

19. Morse, J.W., "Release of Toxic Metals Via Oxidation of Authigenic Pyrite in Resuspended Sediments," in *The Environmental Geochemistry of Sulfide Oxidation,* ACS Symposium Series 550, C.N. Alpers and D.W. Blowes Eds., *Amer. Chem. Soc.* (Washington, D. C. 1994).

20. Rickard, D.T., "Experimental Concentration-time Curves for the Iron(II) Sulphide Precipitation Process in Aqueous Solutions and Their Interpretation," *Chem. Geol.* 78:315 324 (1989).

21. Morse, J.W. and J.C. Cornwell, "Analysis and Distribution of Iron Sulfide Minerals in Recent Anoxic Marine Sediments," *Marine Chem.* 22:55-69 (1987).

22. Cornwell, J.C. and J.W. Morse, "The Characterization of Iron Sulfide Minerals in Marine Sediments," *Marine Chem.* 22:193-206 (1987).

23. Morse, J.W. and T. Arakaki, "Adsorption and Coprecipitation of Divalent Metals with Mackinawite (FeS). *Geochim. Cosmochim. Acta* 57:3635-3640 (1993).

24. Lin, S. and J.W. Morse, "Sulfate Reduction and Iron Sulfide Mineral Formation in Gulf of Mexico Anoxic Sediments," *Amer. J. Sci.* 291:55-89 (1991).

25. Aller, R.C., "Diagenetic Processes Near the Sediment-water Interface of Long Island Sound. I. Decomposition and Nutrient Element Geochemistry (S, N, P)," *Adv. Geophys.* 22:237-350 (1980).

9

EFFECTS OF BIOTURBATION ON SOLUTE AND PARTICLE TRANSPORT IN SEDIMENTS

Gerald Matisoff
Department of Geological Sciences
Case Western Reserve University
Cleveland, OH 44106

1. INTRODUCTION

Nearly all recent fine-grained sediment deposits are inhabited by infaunal macroinvertebrates. These animals, through their burrowing, feeding, excretion, respiration, and locomotion activities can significantly affect the physical and chemical properties of sediments in a variety of ways. Benthos can alter sediment fabric, texture, porosity, and compaction; bind sediment particles and fabricate burrows that are sometimes maintained with semi-permanent polysaccharide linings; mix sediment particles; selectively feed on certain sizes or classes of sediment; secrete mucus and excrete metabolites; pump pore fluids and overlying water through their burrows and through the sediment; alter oxidation/reduction conditions; affect the chemical exchange between sediments and overlying water; and alter the early chemical diagenesis of sediments. Because of this wide variety of mechanisms in which benthos interact with their physical and chemical environment, pollutants such as metals, nutrients, and xenobiotics which are buried in sediments cannot be assumed to forever remain buried. It is therefore essential to consider the macrobenthos in all sediment contaminant studies and in the establishment of any sediment quality criteria.

It is not possible in one chapter to review all the literature that details the entire scope of animal-sediment relations. Excellent reviews by different authors on various aspects of the problem are presented in McCall and Tevesz [1]. More pertinent to this book are studies of how macrobenthos affect the distribution of metals in sediments. Studies of this type are limited and so a detailed review of that work is not particularly useful. The approach taken in this chapter, then, is to provide an overview of the range of ways that macrobenthos affect particle and solute transport in sediments, especially with respect to the distribution and mobilization of metals. It is hoped that the material reviewed here will be comprehensive enough for the non-specialist to understand the current state of research in this area and to have an adequate reference list from which to begin more detailed investigation. Specifically, how animals of different functional types interact with bottom sediments to induce changes will be presented. Then, field and laboratory data and mathematical models of organism effects on solute fluxes, sediment mixing, oxidation/reduction conditions and pore fluid concentrations will be discussed. Finally, a relatively new laboratory procedure to determine the effects of benthos on particle and solute transport will be demonstrated to illustrate the procedure which can be used to quantify the mixing processes.

1.1   Macrobenthos

In addition to infaunal macrobenthos, most modern sedimentary environments are subject to a variety of post-depositional processes which serve to re-mobilize and alter the sedimentary deposit before permanent burial. These processes are illustrated schematically in Figure 1. The top portion of the sediment may be disturbed by wave action. Bacterial degradation of organic matter and chemical reactions change the nature of the sediment and create chemical gradients which induce solute transport through the pore fluids. Invertebrate macrobenthic organisms modify sedimentary processes in numerous ways, as indicated above. Their biological activities result in an increased exchange of overlying water and solutes with those in the sediments and in the redistribution of chemicals within the sediment deposit and of the sediment particles themselves. Altogether, these processes are termed "bioturbation".

1.2   Freshwater benthos

The effects of freshwater benthos on particle and solute transport in sediments has been reviewed by Fisher [2]. Figure 2 illustrates that the profundal macrobenthic community of lakes is numerically dominated by oligochaetes, amphipods, insect larvae (e.g., chironomids and mayflies), and sphaerid and unionid clams [3-5]. Shallow infaunal deposit feeders include amphipods and sphaerid clams which affect only the upper 1-3 cm of the sediment surface [6], but deeper burrowing infaunal forms such as chironomid and mayfly larvae construct semi-permanent burrows (Figure 3) and can significantly alter oxidation-reduction conditions in the sediment to depths of 20 cm or more [7-12]. Infaunal subsurface deposit feeders, such as oligochaetes (Figure 3), and deep infaunal, mobile suspension feeders, such as unionid clams (Figure 3), are capable of thoroughly mixing sediments within their life zones (~10 cm) [13-17]. Boyer *et al.* [18] observed trenches constructed by burbot (*Lota lota*) in the deep waters of Lake Superior which were U- to V-shaped, 3-5 m long, 15 cm wide, and 10-25 cm deep. They conclude that the fish is capable of mixing surface sediment to a depth of 30 cm.

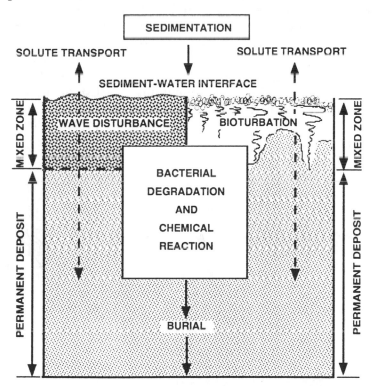

Figure 1.   *Major Processes Affecting the Transport and Fate of Contaminants in Sediments.*

In freshwater environments, tubificid oligochaetes are important agents of bioturbation by virtue of their life habit, widespread distribution, and high population densities [19]. The particle reworking activity of tubificid oligochaetes is highly directional. Tubificids are "conveyor belt" feeders, ingesting reduced sediment at depth in the substratum and expelling this material as mucus-bound fecal pellets at the oxic sediment-water interface (Figure 3). This pelletization creates a high porosity surface layer which enhances solute diffusion. McCall and Fisher [19] determined that the flux of chloride across the sediment-water interface was increased by a factor of 2 in the presence of 100,000 oligochaetes/m^2 (*Tubifex tubifex*). Krezoski *et al.* [14] studied the migration of ^{22}Na in microcosms with the oligochaete, *Stylodrilus heringianus* (67,000 individuals/m^2) and the amphipod, *Pontoporeia hoyi* (47,000 individuals/m^2) and found that the apparent diffusivity was enhanced by about 50% by both animals. In a study of the effects of tubificids on sediment oxygen consumption, McCall and Fisher [19] found that the oxygen consumption of tubificids (*T. tubifex*, 100,000 individuals/m^2) exceeded by a factor of ~2 the simple sum of tubificid respiration and the sediment oxygen demand in the absence of the worms. They determined that about 20% of the enhanced demand was the result of tubificid respiration, while 50-70% of the enhanced demand was due to the oxidation of FeS brought to the sediment-water interface by conveyer-belt feeding. The remaining 10-30% of the enhanced demand was apparently due to increased microbial activity. The actions of tubificids have also been shown to mitigate profound down-core changes in pH [20].

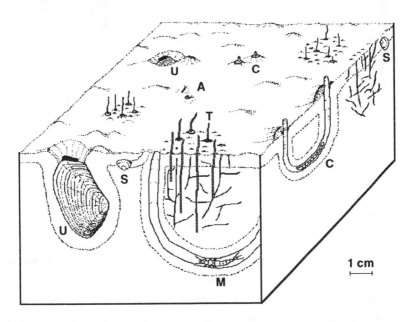

Figure 2.   *Life Positions of Dominant Profundal Freshwater Macrobenthos. A, amphipods; C, chironomids; M, mayflies; S, sphaerid clams; T, tubificid oligochaetes; U, unionid clams. (Modified from [2]).*

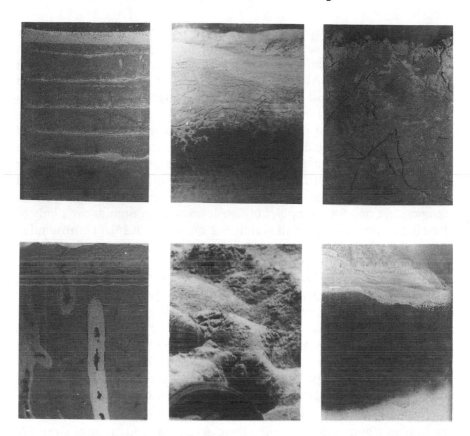

Figure 3.   *Photographs of the Bioturbated Zone of Microcosms Containing Freshwater or Marine Benthos.   Top left, control; Top center, with tubificid oligochaete Limnodrilus; Top right, with tubificid oligochaete Branchiura; Bottom left, with chironomid larvae Chironomus plumosus; Bottom center, with unionid clam Lampsilis radiata; Bottom right, with marine clam Yoldia limatula. Note (1) horizontal color banding caused by oxidation of the sediment-water interface during preparation of added tracer layers; (2) color (chemical zonation) is strongly vertically stratified in the absence of benthos but is partially obscured in the presence of any benthos and is around vertically-oriented chironomid burrows; (3) penetration of oxygenated chironomid burrows deep into reduced sediment; (4) surface pelletal layer in the top 0.5-1 cm in the presence of either specie of tubificid oligochaete; (5) 2-4 cm deep furrows made by unionid clams during movement. Scale bar = 2 cm.*

Advective transport of pore and/or overlying water across the sediment-water interface is potentially the most significant way in which macrobenthos might affect chemical diagenesis and materials balance in freshwater systems. Although nearly all benthic organisms are capable of causing some fluid advection, infaunal organisms that engage in active burrow irrigation, such as chironomids and burrowing mayflies [21,22] or that might inject water to aid burrowing, such as unionid clams [13], are probably the largest contributors to biogenic fluid advection. Chironomid larvae (especially the genus *Chironomus*) are particularly abundant in the profundal zone of lakes. The size and weight of the larvae depends upon a variety of factors such as temperature and food supply, but fourth-instar chironomids are likely to be 10-22 mm in length and weigh 1-2 mg dry [23-26]. Chironomids burrow into bottom sediments; most larvae live in the upper 8-10 cm of the substratum, with occasional occurrences at 40-50 cm depths [27-29]. Most chironomids line their burrows with a transparent, fibrous salivary secretion within which they filter feed by drawing in water by means of anterioposterior undulations of its body through a net spun in the lumen of the tube [22,30]. It is also thought that chironomids perform as surface deposit feeders, scraping the top 1-2 mm of sediment around their burrows for detritus and bacteria attached to fine-grained sediment particles. The larvae are mobile, leaving the bottom (usually at night) and undergoing migration up to 1 m into the water column [31].

The effects of chironomids on the flux of materials across the sediment-water interface have received the most attention [7-9,11,12,32-38]. Ganapti [32], Edwards [33], and Edwards and Rolley [7] demonstrated that chironomids significantly enhance the flux of solutes across the sediment water interface and achieve porewater concentrations equivalent to those in the overlying water within a few hours. Tessenow [34] demonstrated that *Chironomus plumosus* significantly enhances the flux of silica from sediments into the overlying water. Matisoff *et al.* [11] determined that silica fluxes were enhanced by factor of 2, 3, and >3 with *C. plumosus* densities of about 1100, 3300, and 6800 individuals/m^2. Graneli [9] also measured enhanced silica fluxes ($\times 2$-$\times 3$; 1000 individuals/m^2) as well as enhanced phosphorus ($\times 2$-$\times 4$); ammonia ($\times 0.6$-$\times 155$); iron ($\times 1$-$\times 14$); and manganese ($\times 2$-$\times 4$) fluxes. Gallepp *et al.* [36] and Gallepp [37] reported lineally increasing phosphorus fluxes up to a factor of 3 using *C. riparius* and *C. tentans* densities up to 6585 individuals/m^2. The importance of chironomids in pumping oxygenated overlying water into reduced sediments can be seen in Figure 3. Chironomid burrows act as conduits for solute exchange with the overlying water, creating a 3-dimensional mosaic of biogeochemically zoned microenvironments rather than a vertically stratified distribution. Edwards and Rolley [7], Hargrave [8], Anderson [39], and Graneli [9] all reported that the oxygen demand in sediments inhabited by various chironomid species was as much as twice the simple sum of chironomid respiration and sediment oxygen demand in the absence of chironomids. Some experimental results of solute transport influenced by chironomids will be presented below.

The feeding and metabolic activities of burrowing mayfly nymphs and chironomids are functionally similar, so enhanced solute transport by mayflies would also be expected. However, there are few studies of the effects of *Hexagenia* on solute transport. Lawrence *et al.* [40] did conduct laboratory microcosm studies using *Hexagenia* and found that the insect significantly increased the mineralization of sediment nitrogen and the flux of ammonia across the sediment water interface. All other studies have focused on the effects of overlying water chemistry (especially dissolved oxygen) on the activity and survival of the insect [41-45] and on the bioaccumulation of metals and toxics from the sediment [46-49]. These bioaccumulation studies have shown that mayflies are significant vectors for the direct transmission of sediment contaminants such as PCBs, Cd and Hg to fish. *Hexagenia* nymphs are extremely sensitive to pollution themselves (particularly low dissolved oxygen). For example, their abundance in western Lake Erie declined catastrophically from dominance prior to the early 1950s to being absent by 1960. Conversely, over the same period oligochaetes, chironomids, and sphaerids increased in abundance. However, the recent changes in limnological conditions of western Lake Erie from eutrophic to mesotrophic status will most likely result in a recovery of the benthic community and the return of *Hexagenia* as a dominant organism in the near future [3]. In fact, mayfly larvae are beginning to repopulate areas of the Great Lakes [50] and have been found in Lake St. Clair, the St. Mary's River, and the Detroit River [47,51-53]. It is unclear what the effects of the recent, explosive, invasion by the zebra mussel, *Dreissena polymorpha*, will be, except that its filter feeding will likely improve oxygen conditions and promote mayfly recolonization. *Hexagenia* is a vigorous burrower and an important food source for fish in the Great Lakes, and since it lives and feeds in the sediment, its burrowing activity and bioaccumulation of metals and toxic organics may result in the release of pollutants to the water and transport up the food chain.

Just as tubificid and chironomid burrowing and feeding activities alter a number of physical and chemical properties of cohesive sediments, freshwater bivalves also directly alter a number of chemical properties of the sediment porewater and indirectly control the flux of dissolved solutes across the sediment-water interface. Most clams are large by comparison to tubificids and chironomids, and therefore some effects might be expected to be magnified in sediments inhabited by them. Studies using freshwater bivalves are limited, but McCall *et al.* [17] found a 4-fold increase compared to control microcosms in the flux of inorganic nitrogen across the sediment-water interface in laboratory microcosms with unionid bivalves (*Lampsilis radiata*). They attribute about 10% of the increase to radial diffusion through the burrow wall (Figure 3), about 50% of the increase to excretion by the clams, and the remainder to enhanced bacterial metabolic activity near the burrow wall. However, in many North American freshwaters the unionids are not particularly abundant, and the recent introduction of the zebra mussel (*Dreissena polymorpha*) has resulted in competition sufficiently severe to cause the apparent demise of the unionid population throughout the Great Lakes [54] and presumably in many other freshwater settings as well.

The Sphaeriidae (also known as Pisidiidae) are fairly small bivalves (2-20 mm length; 0.5-2 mg dry weight) which occur in densities ranging from a few hundred to a few thousand individuals per square meter in profundal lake sediments [25,55-58]. Sphaerids occur over a wide range of depths. For example, sphaerid densities in Lake Michigan reach their maximum density at 40-50 m and occur to depths of 225 m [59]. Many sphaerids are filter feeders on detritus, bacteria, algae, and protists [60,61], but deposit feeding has also been reported [60,62]. Different life positions have been reported (Figure 2) with significant implications for sediment-water chemical exchange, but this has not been investigated. Similarly, studies on the effects of amphipods and nematodes on the chemical exchange of solutes between sediments and water have not been conducted.

1.3   Marine benthos

The marine benthic community is more complex than the freshwater system (Figure 4). McCall [63], Pearson and Rosenberg [64], Rhoads *et al.* [65,66], and Rhoads and Boyer [67] have demonstrated that the marine benthic community in muddy bottoms tends to follow a successional pattern of pioneering species followed by equilibrium species. They observed that small, opportunistic tube-dwelling polychaetes are the first fauna to colonize a disturbed bottom, and that tubicolous amphipods follow the polychaete colonization. Grassle and Grassel [68], McCall [63,69], and Zajac and Whitlach [70] have identified an opportunistic assemblage in muddy subtidal sediments in shallow embayments of the Northeastern U.S. characterized by large numbers of small usually near-surface dwelling spionid polychaetes (*Streblospio benedicti, Polydora ligni, Polydora cornuta*), capitellid polychaetes (*Capitella capitata, Heteromastus filiformis*), and the amphipod *Ampelisca abdita*. Most pioneering organisms tend to be tubicolous and/or sedentary, live near the sediment-water interface, and feed near the sediment surface or within the water column. Higher-order successional fauna are found when bottom disturbances are infrequent. A late successional stage is dominated by infaunal deposit feeders, many of which are conveyor-belt style feeders, although some polychaetes can switch from deposit feeding to passive suspension feeding at high suspended particle fluxes [71]. Some of these species are also tubicolous (e.g., Maldanidae). Equilibrium assemblages are associated with a deeply oxygenated sediment surface where the "redox potential discontinuity" commonly reaches a depth of over 10 cm and transfer of both water and particles over vertical distances of more than 20 cm is common. Some animals, such as callianassid shrimp routinely burrow to depths of 1 m and have been observed at depths >3 m. Sanders [72-74] and Rhoads and Young [75] described a late successional assemblage of large, deep dwelling deposit-feeders characterized by the polychaete *Nephtys incisa*, and the bivalves *Nucula annulata* and *Yoldia limatula* (Figure 4) that was dominant in subtidal muds from Buzzards Bay, Massachusetts to Long Island Sound. Maldanid polychaetes of several species (*Maldanopsis, Maldane, Owenia*, and others) were also abundant.

The effects of marine benthos on the physical properties of marine sediments have been reviewed by Rhoads and Boyer [67] and on particle redistribution and solute exchange between sediments and water have been reviewed by Aller [76]. Some more recent work has addressed particle selective feeding by benthos [15]. Matisoff and Robbins [77] demonstrated that preferential feeding on tracer-laden particles can retain the tracer within the feeding layer indefinitely. Rejection of large particles during feeding has been identified as the cause for the formation of a surficial gravel layer by McCave [78], and Wheatcroft [79] determined that finer particles penetrated deeper into sediments than larger particles. Particle selective feeding is also related to particle density and food quality [80-82].

Studies of the redistribution of sediment particles have most commonly involved the interpretation of the down core distribution of radionuclides in sediments. Goldberg and Koide [83] developed the first diffusion model to explain observed homogeneity in radioisotopes in pelagic sediments. Since the publication of their model, a large number of diffusion models have been applied to observed distributions of many different radionuclides in sediments to "unravel" the effects of particle mixing to determine sedimentation rates in mixed sediments.

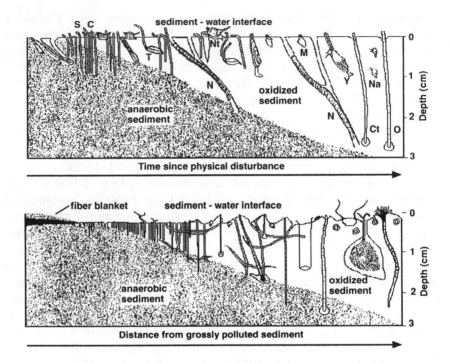

Figure 4.   *Marine Benthic Community Development with Time*
               Following Disturbance (top) and Along a Pollution
               Gradient (bottom).   Small, opportunistic tube-dwelling
               polychaetes that feed near the sediment surface or within
               the water column are the first to colonize (left), while
               higher order successional fauna are larger and tend to be
               dominated by infaunal deposit feeders that cause intensive
               particle mixing (right). Typical species are: C, Capitella
               capitata; Ct, Clymenella torquata; M, Mulinia lateralis; Na,
               Nucula annulata; N, Nereis sp.; Nt, Nassarius trivitatuo;
               O, Owenia fusiformis; S, Streblospio benedicti; T, Tellina
               tenuis; Y, Yoldia limatula. Top figure from Rhoads et al.,
               (1978); bottom figure from Pearson and Rosenberg (1976)
               with similar functional types, but European species.
               (Modified from Rhoads and Boyer [67]).

These models have been reviewed by Matisoff [84] and Van
Cappellen *et al.* [85] and more recent extensions have been developed
by Robbins [16] and Fukumori *et al.* [86]. Although the general model
has been mostly applied to ^{137}Cs and ^{210}Pb distributions in freshwater
sediments, it has been used in marine systems for ^{210}Pb distributions by
Bruland [87], Turekian *et al.* [88], Peng *et al.* [89], and Santschi *et al.*
[90]; for [239,240]Pu distributions by Guinasso and Schink [91],

Benninger *et al.* [92], Santschi *et al.* [90], Olsen *et al.* [93], and Schink and Guinasso [94]; for ^{234}Th distributions by Turekian *et al.* [88], Aller *et al.* [95], Cochran and Aller [96], Demaster *et al.* [97], Santschi *et al.* [90], and Martin and Sayles [98]; for ^{7}Be by Krishnaswami *et al.* [99]; for ^{137}Cs and other artificial radionuclides by Olsen *et al.* [93], Cutshall *et al.* [100], and Anderson *et al.* [101]. Chernobyl fallout over much of Europe has given ^{137}Cs dating a new time marker, although most of that sediment tracing has been in freshwater systems [102-108], but there has been no detectable fallout from Chernobyl in North America. Invariably, studies which utilize environmental radionuclides in sediments do not include simultaneous measurements of the infaunal benthos, so the bioturbation interpretations are tentative, at best.

The activities of benthos not only redistribute sediment particles but also influence the exchange of water and solutes across the sediment-water interface and Aller [76] has reviewed these processes. Burrows act as channels for the direct communication between interstitial and overlying waters (irrigation) [109-112]. Fecal pellets deposited on the sediment surface increase the porosity of the uppermost sediment and therefore increase the diffusional exchange [113]; burrows act as sites for enhanced microbial activity [114]; and the injection of surface water for burrowing and feeding modifies the porosity and chemical conditions in the sediment and increases the exchange of water and solutes across the sediment-water interface [115-117]. Mathematical models of the exchange of pore fluids in the absence of benthos are well developed [118]. These are most commonly one-dimensional diffusion models of an individual aqueous specie, but they have also been coupled to sequential oxidation/reduction reactions and mineral precipitation reactions. These models have also been extrapolated for use in the presence of macroinfauna, usually by incorporation of an apparent diffusion coefficient in the bioturbated zone that is elevated over that in the bulk sediment [19,94,119-125] and this approach will be presented in more detail below. Hammond and Fuller [116] and McCaffrey *et al.* [117] described the exchange of solutes between well-mixed and distinct reservoirs of sediment and overlying water as an advection velocity. Permanent or semi-permanent tube structures or significant directed exchange of water cannot necessarily be described by simple one-dimensional models or by quantitatively modifying a molecular diffusivity. Aller [126,127] has developed a cylindrical burrow model in which vertical diffusion to the sediment surface and radial diffusion to a centrally-irrigated burrow occurs. Diffusion through semi-permanent polysaccharide burrow linings [128] also affects solute transport. The advantage of this model is that is does not require empirically elevated diffusion coefficients or advection velocities. This model will be demonstrated below. This process has also been described in terms of a nonlocalized exchange parameter in which the time rate of change of concentration of a porewater solute due to irrigation is proportional to the concentration difference between porewaters and overlying waters [129-132].

The marine polychaete *Nereis sp.* has been shown to increase the flux of phosphorus from sediments by a factor of 2-6 over that calculated assuming Fickian diffusion [133,134]. Aller and Yingst [124] examined changes in porewater, sediment, and overlying water chemistry caused by the marine bivalve *Yoldia limatula*. They found increases in the NH_4^+ and Mn fluxes from sediment with *Yoldia* which they attributed to both enhanced diffusion and increases in the rate of microbial metabolic activity. Henriksen *et al.* [135] studied the burrow environments of the marine bivalves *Mya arenaria* and *Macoma balthica*. They observed marked increases in the flux of inorganic nitrogen (mostly NH_4^+) from the sediment. They determined that the rates of ammonium production, nitrification, and denitrification were increased within the burrow environment.

2.   EXAMPLE STUDIES

2.1   Solute transport

Microcosms with and without benthos have been used successfully to determine the exchange of solutes between sediment and overlying water [11]. A typical microcosm is shown in Figure 5. It was constructed mostly of acrylic and consisted of four main components: (1) an overlying water reservoir of about 5 liters that contained a magnetically mounted stirring mechanism and could be sealed to simulate the gradual onset of anoxia; (2) a core tube which contained the sediment column (either homogenized mud or a field core); (3) an extruder base which is connected to a pressurized line; and (4) an extruder piston which is installed in the bottom of the core tube and which is used when analyzing the sediment. When a microcosm was disassembled for analysis, the overlying water reservoir was emptied and detached from the sediment chamber and extruder base assembly. After connection of the base to the pressurized line, the extruder piston was forced upwards and the desired intervals of sediment were collected by means of a slicer fitted to the top of the sediment core tube. Porewaters were expressed from the sediments using squeezing in a N_2-filled box.

Tubificids (10^5 *Limnodrilus spp.*/m^2) were added to ten of the microcosms and ten additional microcosms served as controls (T = 15 °C). After 30 days of bioturbation all microcosms were sealed, and the concentrations of nutrients and other materials in the overlying water were monitored. At intervals, two microcosms (one with worms and one without) were sectioned to determine porewater concentrations. In a similar experiment, fourth instar chironomids (*C. plumosus*) were added to three microcosms (1096, 3289, and 6798 individuals/m^2) and a fourth microcosm served as a control. In contrast to the tubificid experiment, the overlying water was aerated during the entire experiment. A third, related experiment utilized the unionid clam *Lampsilis radiata* (92 individuals/m^2), but because of their large size, a regular 10 gallon fish aquarium was used instead of the microcosm shown in Figure 5. Sediment porewaters were obtained by coring the aquarium, sectioning the mud, and centrifuging to separate the interstitial water.

Photographs of microcosms containing each of these experimental animals as well as those used in the particle mixing and solute transport experiments presented in a later part of this chapter are shown in Figure 3. The control core illustrates the classical formation of a thin (~0.5 cm) surficial oxidized layer (reddish in color) overlying a more reducing sediment (grey). The reducing zone also shows microenvironments of different reducing and/or diagenetic mineralogy (black spots). Oligochaetes are found in the top 20 cm of sediment, but feed primarily in the top 2-8 cm [136]. A thin (~0.2 cm) pelletized layer with small (~0.5 cm diameter) fecal pellet mounds is evident in the *Limnodrilus spp.* microcosm but the sediment appears more extensively reworked in the *Branchiura sowerbyi* cell. The oxidized zones are more diffuse compared to the control, and the varying reducing conditions and diagenetic mineralogy is more pronounced as indicated by the mottled appearance of the reducing zone. This mottled appearance is very apparent in the *Branchiura* microcosm. Note that there is no color change around tubificid burrows implying that they are not conduits for active exchange of overlying water. In contrast, the chironomid larvae established well irrigated burrows in the upper 8-10 cm of the substratum. Aller [76] predicts that chemical zonation around an irrigated burrow should occur due to the sequential consumption of O_2, NO_3^-, $Fe(OH)_3$, and SO_4^{2-} during organic matter decomposition. This reaction zonation can be clearly seen around the chironomid burrow. Although not shown, the freshwater unionid clams also cause chemical zonation around their burrows (the photograph shown here only illustrates the furrows caused by their movement). The marine clam, *Yoldia limatula*, created a 2 cm layer of stratified reworked sediment over the reduced, unmixed sediment. The lighter material is "oxidized" and attains its color from ferric oxyhydroxides, while the darker material attains its color from reduced compounds, especially FeS. It is clear from the photographs that the presence of an irrigated burrow enables oxygenated overlying water to penetrate 6 to 8 cm into a zone that was once reduced sediment. The burrow introduces a three-dimensional geometry in the sediment and induces three-dimensional solute transport. Oxygen diffuses into the sediment from the sides of the burrow as well as from the top surface. Unlike the chironomids, the bivalves play a passive role. Oxygen diffusion will occur as long as the burrows are maintained; it does not require the direct injection of oxygenated overlying water by the clam.

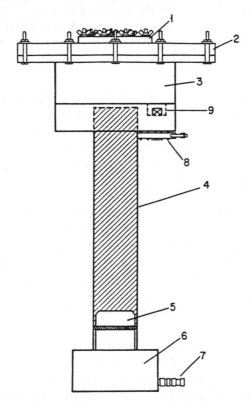

Figure 5.   *Laboratory Microcosm.   1. Sampling cover.   2. Detachable*
               top.   3. Overlying water reservoir (5 L).   4. Sediment
               chamber (core tube 7.6 cm dia ×50 cm long).   5. Extruder
               piston.   6. Extruder base.   7. Quick connect valve.   8.
               Turbine magnetic stirrer.   9. Well-mounted stirbar.

2.2   Overlying water

Typical concentration versus time plots for ammonium, alkalinity, pH, soluble reactive phosphorus (SRP), soluble reactive silicate (SRS), and ferrous iron in the water overlying the oligochaete microcosms are given in Figure 6. The overlying water, which was initially deionized, quickly reached an apparent steady state composition with respect to both SRP and alkalinity.

After the microcosms were sealed and purged with helium (+44 days), alkalinity in the overlying water increased markedly. In most cases alkalinity reached higher values in microcosms with tubificid populations (not apparent in these cores). It is possible that this occurs because of enhanced organic matter decomposition within the pelletal zone or because of enhanced diffusion through the pelleted surface. Immediately prior to (or concurrent with) the rapid increase in ferrous iron in the overlying water (~110 days), alkalinity increased slightly and pH decreased slightly because of the release of H^+ during ferric hydroxide reduction. In natural lake cores the increase in alkalinity is less than in the tubificid core tubes. The exact time at which dissolved oxygen in the overlying water was totally depleted was uncertain. Electrode measurements of dissolved oxygen indicated no dissolved oxygen present prior to other chemical changes (e.g., increase in ferrous iron concentration) known to accompany anoxia. Consequently, the presence of a detectable level of ferrous iron was taken to indicate anoxia in the overlying water. Ammonia was first detected in the overlying water after the microcosms were sealed and helium purged. In general, the concentration of ammonia increased with time. Microcosms with tubificid populations exhibited higher ammonia concentrations compared to those lacking a worm population. Silica concentrations increased uniformly with time and appeared to be unrelated to anoxia, in contrast to Mortimer's [137] results. Microcosms with tubificids exhibited slightly higher silica concentrations than the controls. This may be due to an enhanced diffusive flux in the pelletal layer [19].

In the chironomid cores, the initial overlying water was 1:1 (v:v) deionized and aged tap water mixture. Hence, the carbonate alkalinity (Figure 7) has an initially high background concentration. Microcosms containing chironomids exhibited higher alkalinities than the control (not shown). This effect seems to be real, because the higher the chironomid density, the higher the alkalinity (population density data not shown). It is probably due to an additional diffusive flux from the sediment into the burrows [95,135]. The flux of silica from microcosms containing 6798 chironomids/m^2 is about a factor of 3 greater than that of the control microcosms (Table 1, direct flux measurements), while proportionately lesser fluxes occurred when there were lesser numbers of larvae. The role of animals in enhancing the flux of dissolved solutes across the sediment-water interface is clearly illustrated in Figure 8, where, assuming a simple 2-layer model, the apparent diffusion coefficient for chloride in the bioturbated zone is about a factor of two greater in the presence of unionid clams than in the control.

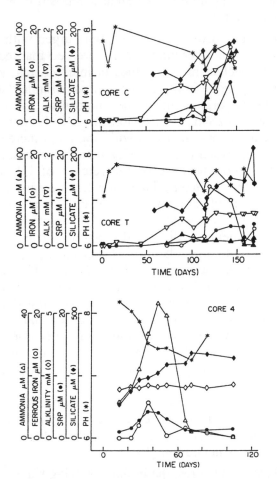

Figure 6. *Concentration of Solutes with Time in the Overlying Water of Microcosms Containing No Benthos (Core C, top), Tubificid Oligochaetes (Core T, middle), and a Field-collected Core (Core 4, bottom). The control and tubificid microcosms were sealed on day 44 and allowed to gradually go anoxic and the natural cores were sealed immediately.*

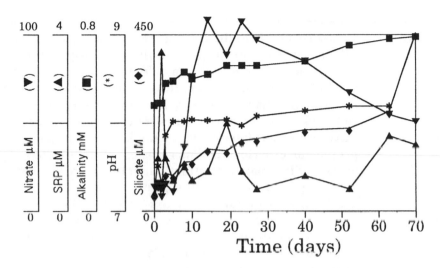

Figure 7. *Concentration of Solutes with Time in the Overlying Water of Microcosms Containing 6798 Chironomids/m². Unlike the tubificid experiment shown in Figure 6, the overlying water was oxidized for the entire duration of the experiment.*

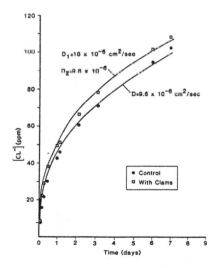

Figure 8. *Apparent Diffusion Coefficients for Chloride in the Presence of Unionid Clams. Assuming a simple 2-layer system of bioturbated sediment overlying unbioturbated sediment the clams appear to enhance solute transport by about a factor of 2.*

Table 1. Comparison of Estimates of the Silica Flux from Microcosms Containing Tubificid Oligochaetes, Chironomid Larvae, and Unionid Bivalves, and from Field Measurements. Direct fluxes were calculated from the accumulation of solutes in water overlying field collected cores or experimental microcosms. Indirect fluxes were calculated using Fick's law from porewater profiles of experimental microcosms or field collected cores. Flux boxes (bottom chambers) encapsulate overlying water and a small surface area of sediment for a brief period (usually 24 hours or less), allowing the build-up or depletion of solutes in the overlying water. Standard deviations based on 6 samples in the oligochaete experiment and 5 samples in the laboratory measurement of intact field-collected cores.

	Direct Flux 10^{-6} mol/m^2/d	Indirect Flux 10^{-6} mol/m^2/d
Tubificid Oligochaetes		
(0/m^2)	1216 ± 610	3466 ± 926
(100,000/m^2)	825 ± 384	2534 ± 761
Chironomid Larvae		
(0/m^2)	1393	1651
(1096/m^2)	2644	855
(3289/m^2)	4191	150
(6798/m^2)	4335	76
Unionid Bivalves		
(0/m^2)	1087	673
(92/m^2)	810	868
Western Central Lake Erie		
Lab Experiment	3485 ± 871	2317 ± 1985
Field Core		853
Western Lake Erie		
Field Core		520
Field Peeper		548
Flux Box	4107	

2.3   Interstitial water

Microcosms in the tubificid oligochaete experiment were periodically selected for porewater chemical analysis, and the results for the control cores, cores with worms, and natural bottom cores are presented in Figures 9, 10, and 11, respectively. The overlying water became anoxic at about 110 days. Ferrous iron concentration profiles show an increasing trend with depth in microcosms both with and without worms, although there is no evidence for a time dependency. Ferrous iron profiles from microcosms with tubificids are somewhat more chaotic than those from microcosms without worms. It is unclear why the control microcosm did not develop the "typical" subsurface maximum observed in the natural bottom cores, except that it is likely that the experimental cores have not reached equilibrium in the mass redistribution of iron due to iron hydroxide reduction reactions of newly buried sediment and oxidation of upwardly diffusing ferrous iron. Kikuchi and Kurihara [138], in a study of tubificids on the chemical and physical characteristics of submerged rice field soil, found that the presence of tubificids *(L. socialis* and *Branchiura sowerbyi*, 8842 indiv./m^2) kept the activity of Fe^{2+} high in the upper 1 cm of the sediment and increased the movement of Fe^{2+} into the overlying water. In general, the data for phosphate indicates that it behaves similarly to iron. Iron and phosphate in lake sediments are often correlated positively by adsorption reactions [137] and antithetically by vivianite equilibria [139]. Both controls on phosphate appear to exist in these experiments. In the natural lake cores, the depth of the SRP subsurface maximum coincides with that of the ferrous iron, and the SRP concentration is reasonably constant at depth. The magnitude of the SRP maximum increases up to 66 days, then decreases. This is in contrast to the ferrous iron maximum which consistently increases. Ammonium concentrations usually increased exponentially to a depth of ~10 cm in the control microcosms, and then remained constant. The ammonium concentrations did not exhibit a strong time dependency, but concentrations at depth did tend to increase slightly with time. The absolute value of ammonium at depth was similar to that observed in natural lake cores (~200 μM). In microcosms with tubificids the observed ammonium profiles are strikingly different. Ammonium concentrations usually exhibited far less depth variation and there was a local concentration maximum at a depth of about 4 cm. This subsurface maximum was also observed in the natural lake cores, but there is greater between-core consistency in the natural lake cores. The subsurface maximum indicates that an additional source of ammonia is

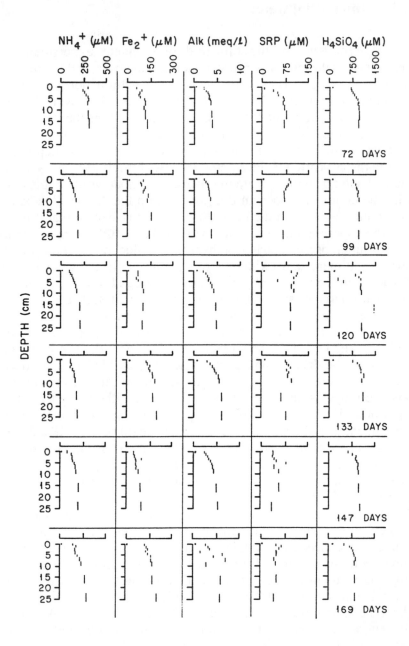

Figure 9.   *Interstitial Water Chemistry from Microcosms Containing No Added Benthos. Six core tubes were extruded and analyzed to obtain the time series data.*

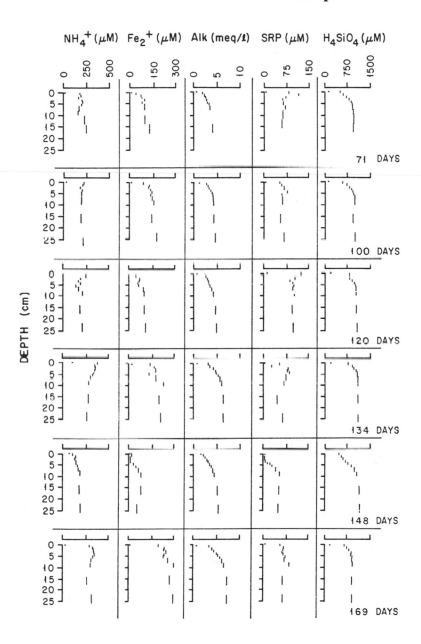

Figure 10. *Interstitial Water Chemistry from Microcosms Containing Tubificid Oligochaetes (10⁵ Limnodrilus spp./m²). Six core tubes were extruded and analyzed to obtain the time series data.*

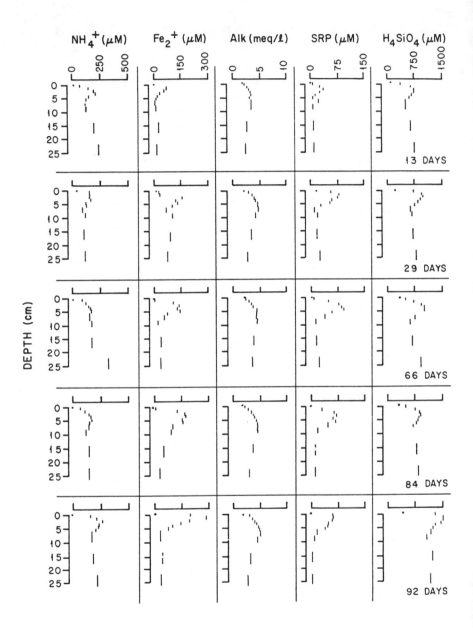

Figure 11. *Interstitial Water Chemistry from Microcosms Containing Field-collected Cores. Five core tubes were extruded and analyzed to obtain the time series data.*

present in the upper 3-5 cm of worm inhabited sediments and strongly supports the interpretation that ammonia is sourced from both organism excretion and enhanced microbial activity. It also suggests that tubificids may be excreting enough ammonia to "buffer" porewaters against changes in pH [20]. Kikuchi and Kurihara [138] also found that tubificids enhanced the ammonium concentration in the sediment and the ammonium flux across the sediment-water interface. Tubificids have also been shown to enhance the consumption of nitrate by sediments. Chatarpaul et al. [140,141] showed that the presence of tubificids increased the rates of both nitrification and denitrification in sediments. They further showed that, even when clean glass beads were used as a substratum for the tubificids, the concentration of nitrate in overlying water decreased. Furthermore, the decrease in nitrate concentration could be stopped by removing the worms. This finding suggests that denitrifying bacteria were associated with the oligochaetes. Indeed, Chatarpaul et al. [141] isolated nitrifying and denitrifying bacteria from the body walls and guts of tubificids. Depth profiles of alkalinity from microcosms, both with and without worms, are very similar. Alkalinity profiles exhibit an exponential increase with depth to ~10 cm, then increase monotonically with depth. In the microcosms with tubificids and in the natural lake cores, alkalinity profiles do exhibit evidence of time dependency. Profile curvature changes with time and the alkalinity at depth increases with time, at least in the sediments with tubificids. Natural lake cores exhibited a subsurface maximum just below the depth at which ferrous iron reaches a constant value. Silica concentrations in microcosms, both with and without tubificids, generally exhibit an increase to a depth of ~5 cm and thereafter maintain a fairly constant concentration, although the silica concentrations at depth tend to increase with time. In the natural lake cores, silica profiles exhibit a subsurface maximum. The concentrations tend to increase with both depth and time below the maximum. The vertical location of the maximum is coincident with those of ferrous iron and SRP, although the reason for the maximum may be unrelated and due instead to the relative quantity of diatom frustules available for dissolution [142,143].

Microcosms containing the chironomids were analyzed 84-88 days after introduction of the larvae. Figure 12 shows that by pumping oxygenated overlying water through their burrows chironomids significantly affect the porewater concentrations of ammonia, silica, iron, and manganese, and the higher the animal density, the more pronounced the effect. At high densities, ferrous iron concentrations in the top 10 cm of the sediment are very low. This is due in part to the large amount of open burrows and the large volume of oxidized sediment around the burrows which contain little dissolved iron. When there are large numbers of organisms, the shorter diffusion path to oxygenated water (into burrows) also inhibits the build-up of reduced compounds. Phosphate (not shown) behaves similarly to iron; most likely its concentration is controlled by adsorption onto ferric oxyhydroxides near the sediment-water interface as well as near the burrow wall-water interface. This illustrates that because of the effects of irrigation, reactions that typically occur at the sediment-water interface can be expected to have increased importance in irrigated sediment. The ammonium profiles are also strikingly different from the controls. Ammonium concentrations are reduced in the top 5-10 cm of the sediment column and increase rapidly below the burrowed zone. This trend, however, is not as strong as seen in the iron or phosphate profiles. This means that the ratio of the rate of supply to the rate of removal of ammonia in that zone is greater than similar ratios of iron or phosphate. The oxidation kinetics of iron and the sorption kinetics of phosphate are rapid. This suggests that ammonification and nitrification are significantly enhanced in irrigated sediments. This agrees well with the findings in the overlying water, where chironomids enhanced the nitrate flux by about a factor of 4 over the control. Chironomids also affect denitrification. Andersen [39] reported that the presence of $C.$ $plumosus$ increased the rate of nitrate consumption by sediments. Nitrate consumption by sediments was strongly correlated with the biomass of $C.$ $plumosus$. The consumption of nitrate was dependent on nitrate concentration in the overlying water and $C.$ $plumosus$ enhanced nitrate consumption by 0.11×10^3 mol/m^2/d/mg wet biomass at nitrate concentrations of 0.14-0.21 mM and by 0.39 mol/m^2/d/mg wet biomass at nitrate concentrations of 0.42-0.71 mM. In order for nitrate concentrations to increase in the overlying water, the supply of nitrate by ammonification and nitrification must exceed this enhanced denitrification. Thus, the presence of an actively irrigated burrow alters porewater chemistry and diffusion, which alters sedimentary microenvironments and thus bacterial metabolic activity and organic matter decomposition, which in turn alters the nitrogen flux. Alkalinity concentrations are nearly constant, at the overlying water value, in the top 6 cm of microcosms containing chironomids (not shown), which is most likely due to radial diffusion around the burrows. Silica concentrations in microcosms, both with and without chironomids generally increase to depths of about 15 cm or more. This increase is most likely due to the dissolution of diatom frustules in the sediment, and perhaps some silica-clay mineral exchange. The presence of chironomids limits the concentration of silica in the top 6 cm of the sediment column. This is probably caused by water pumping activities of chironomids, and associated flushing of burrows and induced radial transport to the burrows. By way of illustration, direct flux measurements (made by measuring changes in the overlying water (Figure 7) and indirect flux estimates (made by applying Fick's law to porewater concentration gradients) for silica are compared in Table 1.

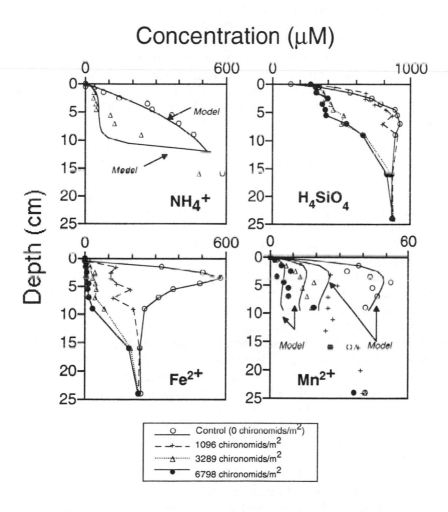

Figure 12. *Interstitial Water Chemistry from Microcosms Containing Various Densities of Chironomids Sampled After 84-88 Days. Also shown are model profiles for ammonium and manganese.*

Direct and indirect flux estimates for silica give comparable results when no chironomids are present, but these two estimates deviate progressively as chironomid population densities increase. At the highest density studied the direct flux estimate is nearly 60 times larger than the indirect flux estimate, because the activities of the chironomids have greatly decreased the silica concentration gradient at the sediment-water interface. Furthermore, the presence of chironomids has increased silica transfer, since the amount of silica that has been contributed to the overlying water is about 4 times greater than can be accounted for by simply flushing interstitial water from the sediments. Burrow irrigation by chironomids maintains a high level of silica unsaturation around the burrow which enhances diatom frustule dissolution.

Also shown in Figure 12 are model results for ammonium and manganese. The model is based on vertical diffusion to the sediment-water interface and radial diffusion to burrows [109,127]. Aller observed that permanent tube dwellings formed by sedimentary benthos produced three-dimensional chemical gradients in the porewaters and solid phases of sediment and increased the flux of material across the sediment-water interface. He conducted laboratory and field studies of *Amphitrite ornata*, a sedentary, surface deposit-feeding polychaete, and *Clymenella torquata*, a sedentary, subsurface deposit feeding polychaete. He found that chemical effects caused by the structure of the tube are the result of the distribution of bacteria around the tube, the way the burrow is maintained, and a change in the geometry of diffusion because of the presence of a hollow tube maintained at seawater concentrations. For ammonium, Aller [109,126,127] considers the burrows to be vertical cylinders that are uniformly spaced and constantly irrigated with overlying oxygenated seawater (Figure 13). He notes that the exact geometry of the burrows is three-dimensionally complex and time-dependent, but that the sediment may be approximated as a mosaic of microenvironments represented on the average by a central irrigated vertical burrow and its immediately surrounding sediment. The ammonium produced by microbial activity diffuses toward areas of low ammonium concentration (burrows and overlying water). Mathematically, he expressed these processes at steady state in terms of cylindrical and vertical diffusion:

$$\left(\frac{D}{r}\right)\left[\frac{\partial(r\partial C/r)}{\partial r}\right] + D\left(\frac{\partial^2 C}{\partial z^2}\right) + R(z) = 0 \qquad (1)$$

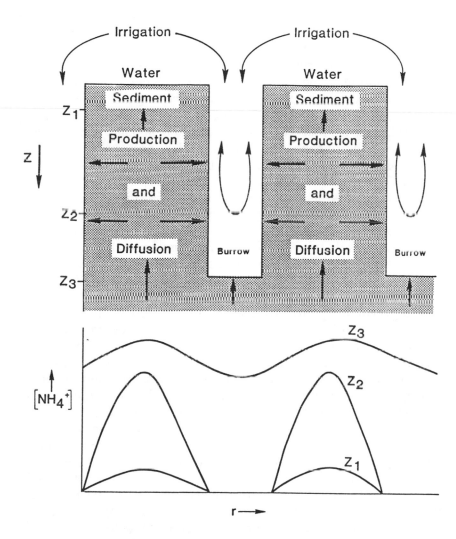

Figure 13. *Schematic Representation of the Cylindrical Burrow Model of Aller [95,109]. Dissolved species generated in the sediment diffuse down the greatest concentration gradient, i.e., radially toward burrows and vertically toward the sediment-water interface. Burrows are assumed to be continuously and instantaneously mixed with overlying water (irrigated). (After Berner [118]).*

where D is the aqueous molecular diffusion coefficient of ammonium in the sediment, r is the radial distance from the nearest burrow axis, C is the ammonium concentration, and $R(z)$ is the rate of microbial production of ammonium. The solution to Equation 1 is shown in Figure 12 applied to the ammonium data from the control and the 3289 chironomids/m^2 experiment. The modeled profile captures the general features of the bioturbated microcosm quite well - uniformly low concentrations of ammonium in the top 6-8 cm increasing up to those of the control at depths below the bottom of the burrows. However, the modeled rate of increase at the base of the burrowed zone is larger than observed. This is probably a result of the model boundary condition at the base of the burrow, since the model is only valid within the burrowed zone and it is poorly coupled to the unbioturbated zone below. The model also reproduces reasonably well the porewater concentrations of Mn within the burrowed zone (Figure 12). Several interesting features are illustrated by the model. First, the larger the burrow density the more Mn concentrations are suppressed. This agrees quite well with the data. At high animal densities, Mn concentrations in the top 10 cm of the sediment are very low. When there are large numbers of organisms, there is a shorter diffusion path of oxygenated water from the burrows into reduced sediment. This results in a large volume of oxidized sediment around the burrows which contain no dissolved manganese, so when that depth interval of sediment is sampled, it yields a low concentration of manganese. Second, and perhaps the most striking feature of both the iron and manganese porewater profiles is the subsurface maximum which is destroyed in the presence of bioturbation. This maximum occurs because of reduction of buried iron and manganese oxyhydroxides and the oxidation of upwardly diffusing reduced iron and manganese. It is more pronounced for iron because oxidation rates of reduced iron and the precipitation of oxidized iron are faster than those of manganese. Burrows disrupt this singular oxidation/reduction band, essentially spreading it out over the entire bioturbated zone. The model simulates these processes quite well, but again, the coupling between the burrowed zone and the unburrowed sediment below is not well characterized. Manganese concentrations with modeled fits in porewaters from natural bottom cores collected from the western and central basins of Lake Erie are shown in Figure 14. Using "typical" chironomid population densities in the two basins, the model describes both the magnitudes of the concentrations and the profile shapes quite well. This illustrates that benthos not only control porewater concentrations of chemicals such as metals, but that their natural bottom population densities can be used to predict the post-depositional behavior of metals in the environment.

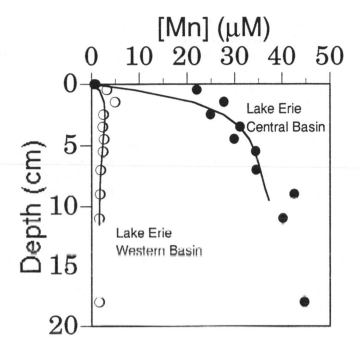

*Figure 14. Manganese Concentrations (data points) with Model
Profiles (solid lines) from Cores Collected from the
Western and Central Basins of Lake Erie. Model fits
derived from chironomid experiment (Figure 13) and
natural bottom population densities.*

2.4   Particle transport

Because many contaminants, including most metals, are associated
with the solid phases in sediments, the fate of these contaminants is
closely tied to the fate of the particulates. Activities by benthic
macroinvertebrates result in the post-depositional mixing of sediments.
Studies of this biological mixing of sediment solids have typically been
one of a few kinds: qualitative or quantitative natural history
descriptions of the activities of important mixing organisms; quantitative
analyses of the effects of mixing of natural tracers in sediment cores; or
laboratory or field measurements of the details of the mixing process
within the biologically active zone of the sediment. Each of these
descriptions of the process of particle mixing is essential, and each has
special strengths and weaknesses.

Most of the early studies were of the first type and have been ably
summarized by Rhoads [144] and Thayer [145]. These studies have

drawn attention to the geologic and geochemical significance of sediment mixing by infaunal macrobenthos, and conceptual models constructed from their results have defined the important problems for future study. Such studies remain important in defining important problems such as new forms of mixing by previously unstudied organisms in poorly studied habitats, and they continue to be performed (e.g., [67,146-155]).

Analyses of the vertical distribution of natural radioisotopes, cosmogenic nuclides, and bomb-produced radioisotopes have allowed the quantification of the intensity and depth of biologic mixing in sediments ranging from intertidal to abyssal sediments over times scales ranging from months to decades (although most of the studies of this type look at mixing over fairly long time intervals). The majority of that work has focused on the calculation of sedimentation rates in the presence of biological mixing (reviewed by Matisoff [84], and Boudreau [156,157] and Boudreau and Imboden [158]) have evaluated the mathematical criteria upon which various models are based. The problem with using only the distribution of natural and bomb-produced tracers in field-collected cores to quantitatively evaluate the sedimentation rate, mixing rate, chemical precipitation/dissolution/ degradation rates, and the solute exchange with the overlying water is that model coefficients are not fully constrained by the radionuclide distributions. This necessitates estimating or arbitrarily fixing the values of some model parameters. One deficiency of these types of studies is that it is not common for the macrobenthos likely responsible for the mixing to be sampled simultaneously with radionuclide distributions. Another is that these studies cannot typically analyze non-steady state mixing dynamics that occur over shorter time intervals that may be of interest in predicting the fate of contaminants in coastal waters [155].

All mathematical models require "calibration" with appropriate field and/or experimental data [158]. Much of the early work of this nature has been summarized [84], and a model has been developed [16] that incorporates many of the mixing processes known to be important [77]. Because there are so many complicated processes that occur simultaneously in the natural setting it is desirable to isolate and quantitatively evaluate individual parts of the system when possible. Experiments which help define the mathematical formulation of some these processes and/or help bound the values of some of the model coefficients are essential to be able to apply these models to predicting the fate of nutrients, metals and toxic substances in sediments. The advantage of laboratory studies of macrobenthic mixing are that non-steady state and species-specific mixing processes are easy to study, and it is possible to model and measure mixing in great detail. A disadvantage of laboratory studies is that rates measured may not reflect *in situ* rates even if the form of mixing is the same. Here, some of the experimental work of this type will be reviewed and an example of a mixing experiment and its data interpretation will be presented in detail in order to illustrate the state of the discipline of particle mixing by benthos.

Only a few quantitative experimental studies which focus on the actual processes of bioturbation have been performed. Using radionuclide tracers the effects of mixing by the major types of Great Lakes freshwater benthos (e.g., [6,14,15,113,159,160]) have been successfully studied. Similar studies of marine systems are more limited and have been mostly conducted at the MERL mesocosm tanks (e.g., [117,161-164]). The majority of the work conducted at the MERL mesocosms has primarily focused on net exchange processes in a scaled-down version of the full system as opposed to isolating individual types of organisms to evaluate specific "feeding functions" [6,113], but it has also demonstrated the utility of experimental studies using radioisotope tracers. Presented here is a laboratory radioisotope tracer study [165] of sediment mixing by *Yoldia limatula*, a common marine bivalve to illustrate how these types of experiments can be used to provide a quantitative description of the depth dependencies of the feeding functions used in many mathematical models of the mixing process.

2.5   Laboratory experiment

Yoldia and sediments for use in the laboratory bioturbation experiments were collected with a PONAR grab sampler from the muddy bottom in Buzzards Bay, Massachusetts. The *Yoldia* samples were size separated into two groups: larger (2.3, 2.5, and 2.7 cm), and smaller (0.7, 0.8, 1.0, 1.0, 1.1, 1.2, 1.3, and 1.8 cm).

Radionuclides well suited for non-destructive tracing of the movement of sediment particles, pore fluids, and/or solutes possess a long half-life (on the order of the length of the experiment or longer) and gamma energy which is sufficiently high to eliminate sediment self-absorption effects. For tracing particle movement, a large adsorption or distribution coefficient is desirable; for tracing solute movement, a small K_d is desirable. ^{137}Cs has been successfully incorporated into the clay size fraction of illite for use as a particle tracer. Distribution coefficients of 5000-7400 have been found by desorption experiments and by following broadening of a line source of labeled sediments in the gamma scan system [6]. Similar experiments have been conducted with vermiculite, attapulgite, and smectite and ^{109}Cd, ^{203}Hg, ^{51}Cr, ^{54}Mn,

^{65}Zn, and ^{59}Fe in sea water. Although there is conflicting evidence about the utility of ^{137}Cs in marine systems because of reported desorption from the particles (e.g., [162,166]) work in our laboratory has found that ^{137}Cs as well as ^{51}Cr, ^{54}Mn, ^{59}Fe, ^{65}Zn, and ^{203}Hg are strongly bound to illite in both freshwater and seawater. We have also used ^{22}Na and others have used ^{54}Mn to monitor solute and pore-water transport [161]. Our results also indicate that ^{109}Cd can be used to trace the movement of loosely particle-bound substances.

Experimental microcosms consisted of rectangular plexiglas cells (30 cm L×15 cm W×3 cm D) containing a 15-20 cm sediment column and 10-15 cm of overlying water. The top 6 cm of the sediment column was formed by the addition of multiple 0.05-0.10 cm thick layers of the ^{137}Cs-labeled clay (~1μCi ^{137}Cs) between 1- 2 cm thick layers of unlabeled sediment. Mixing experiments were run at 20°C. After the sediment compacted for several days, organisms were added to the cells.

The experimental gamma scan system has been in operation in various forms for over ten years [6]. Figure 15 shows how the system presently consists of up to three 2-ft long aquaria which house the experimental microcosms, a collimator/detector, and a gamma spectrometer/counting system. The system consists of a computer-operated precision stepper motor connected to a 2-dimensional precision drive screw upon which the collimator/detector is mounted. This now permits unaided 24-hour scanning and data collection without disturbing any of the microcosms. Positioning within a 160 cm long and 40 cm tall rectangle is reproducible to <0.1 mm and vertical scanning is in millimeter (or less) intervals. For most detector positions, sufficient counts can be accumulated in several minutes and an entire 5 cm wide vertical profile can be obtained in several hours. In these experiments the radioisotope is detected by means of a 2"×2" NaI detector coupled to a multichannel analyzer. Each cell was scanned vertically about once a week for the duration of the experiment which was about 60 days.

2.6   Mathematical models of particle mixing

The quantification of biologic mixing of radioactive particle tracers usually involves the construction of a model of the process. The most popular model is a one dimensional diffusion model introduced by Goldberg and Koide [83] for a constant flux input and further developed by Guinasso and Schink [91] for time dependent fluxes of materials into the bottom. Under the right conditions the intensity of biologic mixing is appropriately quantified as a biologic particle diffusion coefficient [156]. The actual formulations of the sediment system are usually

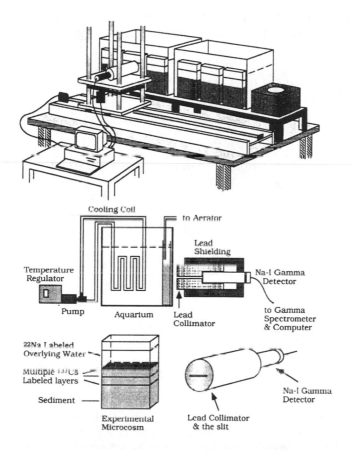

Figure 15.   *Automated Experimental Gamma Scan System.   The top illustration shows the general layout of the equipment. The system consists of up to three 2-ft long aquaria which house the experimental microcosms, a collimator/detector, and a gamma spectrometer/counting system.   The collimator/ detector is mounted on a 2-dimensional precision drive screw which is controlled by a computer-operated precision stepper motor.   The radioisotope is detected by means of a 2"×2" NaI detector coupled to a multichannel analyzer.   Experimental microcosms (bottom illustration) consist of rectangular plexiglas cells (30 cm L×15 cm W×3 cm D) containing a 15-20 cm sediment column with multiple tracer layers between layers of unlabeled sediment and overlain by 10-15 cm of water. The microcosms are contained in a temperature-controlled, glass front 40-liter aquarium equipped with several air bubblers and a dolomite under-gravel filter for adequate aeration and filtration (middle figure).*

advection-diffusion equations, the advection term representing sedimentation and the diffusion term representing biologic mixing, i.e.,

$$\frac{\partial C}{\partial t} = D_b\left(\frac{\partial^2 C}{\partial z^2}\right) - \omega\left(\frac{\partial C}{\partial z}\right) - \kappa C \qquad (2)$$

where

C	=	concentration of contaminant or tracer (g/cm^3 of bulk sediment),
t	=	time (yr),
D_b	=	biologic mixing coefficient (cm^2/yr),
ω	=	sediment accumulation rate (cm/yr),
κ	=	decay constant for the substance (yr^{-1}),
z	=	either depth in the sediment (cm) or mass accumulation in a mass-based coordinate system favored by some authors [167,168].

Modifications of the diffusion analogy, such as the incorporation of depth-dependent biologic diffusion coefficients (such as exponentials or gaussians), have been reviewed by Matisoff [84]. Such models are still the most common and in many cases provide an adequate description of the mixing process. For example, Officer and Lynch [169] have recently used the simple formulation of mixing along with a simple but useful new parameterization of the boundary condition that describes bottom sediment-water column exchange to account for long term (30 year) losses of mercury from contaminated sediments in Bellingham Bay, Washington.

Other forms of mixing are also known to be important in both marine and freshwater environments. The occurrence of advective biologic mixing processes in coastal environments has been recognized for some time [144,170,171]. The dominance of "conveyor belt" mixing by tubificid oligochaetes in mud bottoms of many lacustrine environments [6,19] led to the development of quantitative models of the process [16,113]. In marine environments, conveyor belt feeders that feed at depth and deposit sediment at the sediment-water interface are typically polychaete annelids, although other taxa such as hemichordates, holothurians, and estuarine oligochaetes may also be involved [171,172]. More recently Rice [172] performed a series of laboratory experiments on feeding rates and field observations of four species of the advector *Scoloplos* and [7]Be distributions, and he determined that these polychaetes dominate sediment mixing processes in the shallow cove in which they were studied. Likewise, in stable seagrass and coral reef lagoon carbonate sediments, conveyor belt feeders such as callianassid shrimp and hemichordate worms (enteropneusts) account for most of the sediment turnover [111,173-175]. Other radionuclide tracer evidence suggests the importance of advective mixing in a wide variety of environments in both shallow water [90,163,176,177] and deep [178,179]. Observations of *Yoldia* in these experiments indicates that although the animal is a deposit feeder, it ejects sufficient processed material into the water column that it exhibits both advection and diffusion types of mixing behavior.

Other observations of biota and radionuclide distributions have suggested the local importance of reverse conveyor belt species that exhibit sub-surface defecation [180,181]. Boudreau [157] produced a general model of advective feeding capable of analyzing upward conveyor belt feeding, funnel feeding [182,183], and sub-surface defecation.

The gamma scans of the larger sized *Yoldia* are shown in Figure 16. Four gaussian-shaped peaks indicating the position of each of the four radiolabeled layers are clearly evident at the beginning of the experiment. Robbins *et al.* [6] showed that the response of the collimated gamma detector to a line source of radioactivity is gaussian in form; thus the multiple labeled layers will appear as a series of partly overlapping gaussian distributions. With time several important phenomena occur. First, the peaks become wider. Within about a week there is no longer baseline separation of the peaks. This happens because mixing of sediment causes the tracer from adjacent peaks to be mixed into the unlabeled sediment. The process appears to be slow relative to the time scale of observation and therefore exhibits diffusive behavior. Second, peak heights (areas) relative to background become smaller with time. Since the tracer has a half-life long relative to the length of the experiment, this decrease in activity cannot be attributed to radioactive decay. Feeding causes some of the activity of each peak to be mixed into adjacent unlabeled layers and/or to be deposited at the surface, raising the background activity while lowering the activity of each individual peak relative to the background. Third, the effects of mixing are greater near the surface than at depth. This can be seen by comparing the disappearance of each peak. The peak at the surface is almost gone by 9 days and disappears by 32 days. Similarly, the layer at 2 cm is also destroyed within 32 days. The layer at 4 cm is clearly identifiable at 32 days, but by 50 days it is almost completely mixed. Little change in the layer at 6 cm can be observed in the first week, and by 50 days there is only about a 20% decrease in that peak. These observations indicate that biological mixing by *Yoldia* is greatest near the sediment-water interface and decreases with depth and that it is possible to use the data to determine quantitatively the depth dependencies of the mixing process.

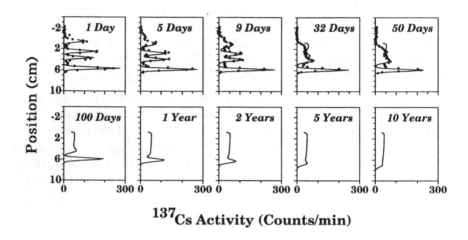

Figure 16. *Observed* ^{137}*Cs Activity for Selected Times in the Medium-sized Yoldia Experiment (data points) and Modeled Profiles (solid lines). Note that biodiffusion results in an increase in peak spread with time, and feeding is detected by a decrease in peak area with time.*

The biological mixing of sediments has been described mathematically as [16]

$$\frac{\partial \hat{C}}{\partial t} = \frac{\partial}{\partial z}\left[D_b \frac{\partial \hat{C}}{\partial z} + \phi D_M \frac{\partial}{\partial z}\left(\alpha\hat{C}\right) \right] - \frac{\partial}{\partial z}\left[\omega\hat{C} + \phi(v - \omega)\alpha\hat{C} \right] - \left[\lambda + \eta\gamma(z) \right]\hat{C}$$

(3a)

$$\text{where} \quad \alpha = \frac{1}{\phi(K+1)}$$

(3b)

$$\text{and} \quad C = \frac{\hat{C}}{\rho_s(1-\phi)}$$

(3c)

where

C = total tracer mass (g/g sediment solids)

$\hat{C}$ = total tracer mass (g/cm^3 bulk sediment),

z = sediment depth (cm) (positive downward),

D_b = coefficient of diffusive mixing of bulk sediment by organisms (cm^2/year),

ϕ = porosity (cm^3 water/ cm^3 bulk sediment) = $\phi(z) = \phi_\infty + (\phi_0 - \phi_\infty)\, e^{-(\beta z)}$,

ϕ_∞ = porosity at depth,

ϕ_0 = porosity at the sediment surface,

β = constant which characterizes the rate of porosity change with depth,

D_m = coefficient for all microscale diffusive processes (cm^2/year),

ω = velocity of sediment solids relative to the interface (cm/year),

v = velocity of porewater relative to the interface (cm/year),

λ = radioactive decay constant for ^{137}Cs ($=0.6932/t_{1/2}$) (year^{-1}),

η = dimensionless contaminant particle selection coefficient on ^{137}Cs by advective feeders,

$\gamma(z)$ = first order depth dependent feeding rate of advectors on bulk sediments (year^{-1}),

K = dimensionless sediment adsorption coefficient for ^{137}Cs,

ρ_s = mean density of the sediment solids (g/cm^3).

Equations in 3 describe the time rate of change at any sediment depth of the total tracer concentration. The first term in brackets on the right hand side of Equation 3 accounts for mixing processes that may be described as diffusive. The first partial differential in that term accounts for diffusive mixing of bulk sediment by organisms while the second partial differential in that term accounts for diffusive transport of the tracer through the sediment porewater and includes the linear adsorption of the tracer between solids and porewater. Technically, D_m also includes solute transport induced by organisms from burrow irrigation, but this effect is ignored in the model. The second term in brackets on the right hand side accounts for tracer transport caused by advective processes. The model accounts for two advective processes: sedimentation and compaction of both pore fluids and solids. The last term on the right hand side of Equation 3 accounts for the loss (or gain) of the tracer at any depth by radioactive decay and advective feeding. Values of η less than one indicate that organisms select against ingestion of the tracer labeled particles while values greater than one indicate that organisms preferentially feed on tracer-associated particles.

It is possible to extract information from the time series of scans to obtain values for the parameters in Equations 3 that describe diffusive mixing, non-local mixing, depth distribution of tracer feeding, and particle selectivity. Small scale "diffusive" mixing can be seen as a change in the spread, σ_i, of each of the layers determined by a nonlinear least squares fit to the data. The widening of the peak and the velocity of burial of the layers under conveyor belt feeding is determined by the repeated measurement of the position of the layers over time. Reworking is also measured by the removal of labeled material from each layer. The tracer feeding rate is in turn related to the rate of particle feeding and velocity of peak burial. Losses of tracer from the feeding zone are accompanied by increases in other layers, which are surface layers in the case of conveyor belt feeders.

Because the three terms in the model are reflected in different ways in the time series scans, it is possible to model different portions of the data to obtain the parameters needed by Equation 3. In the absence of sedimentation and ignoring decay, feeding (peak area), and advection, and assuming that in a narrow depth range the biodiffusion coefficient, D_b, is a constant, the reworking of sediment in the microcosm cells can be approximated by a diffusion model and Equation 3 reduces to

$$\frac{\partial \hat{C}}{\partial t} = D_B \frac{\partial^2 \hat{C}}{\partial z^2} \tag{4}$$

The analytical solution to Equation 4 is given by

$$C(z,t) = \frac{M}{2\sqrt{D_b t}} \exp\left(-\frac{z^2}{4 D_b t}\right) \tag{5}$$

Note that the solution to Equation 4 as shown in Equation 5 is similar to a gaussian:

$$C(z,t) = \frac{M}{\sigma\sqrt{2\pi}} \exp\left(-\frac{(z-\mu)^2}{2\sigma^2}\right) \tag{6}$$

where M is a constant, μ is the position of the peak, and σ is the spread of the peak. From this, it can be seen that $4D_b t = 2\sigma^2$, or

$$\sigma = \left(\sqrt{D_b}\right)\left(\sqrt{2t}\right) \tag{7}$$

By plotting the spread of each peak, σ versus $\sqrt{2t}$, the square root of the particle diffusion coefficient can be determined from the slope of the line. However, because the collimator slit has a certain width (0.4 cm in our system), the response of the detector to the submillimeter radioactive tracer layer is broader than the layer really is. Therefore, all data were corrected to account for this apparent spread by determining the apparent spread at the beginning of the experiment and calculating the spread of a peak at different times using

$$\sigma_t = \sqrt{ABS\left(\sigma_{t,m}^2 - \sigma_o^2\right)} \tag{8}$$

where $\sigma_{t,m}$ is the measured spread and σ_o is the initial spread at t=0 (=0.18 cm). Plots of the least squares fits of σ versus $\sqrt{2t}$ are shown in Figure 17 for each of the four labeled peaks for the larger sized *Yoldia*. Table 2 lists the biogenic diffusion coefficients obtained for each peak for both sizes of *Yoldia*. These results indicate that D_b is about a factor of 5-10 higher at the surface than at a depth of 5-6 cm. This is evident in Figure 17 where it can be clearly seen that the slope of the lines decreases with depth in the core indicating that diffusive mixing by *Yoldia* decreases with depth. This occurs because the animal is mostly a sedentary deposit feeder, removing sediment from around its fixed position just beneath the sediment-water interface and depositing sediment at the sediment surface.

A plot of the slope of the lines in Figure 17 with depth yields the depth dependency of the diffusion coefficient. This mixing behavior may be approximated as a localized gaussian diffusion model of sediment reworking:

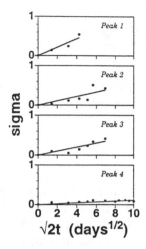

Figure 17.   *Plots of σ versus √2t for Each Peak (Equation 7) for Selected Times in the Larger-sized Yoldia Experiment. Slopes of the lines represent the rate of spreading of an individual peak caused by apparent biodiffusion. Note that the slopes of the lines are greatest near the surface indicating that the diffusion rate decreases with increasing sediment depth.*

$$D_b(z) = D_{b_{max}} \exp\left(-(z-\mu)^2 / 2\sigma^2\right) \tag{9}$$

This is in contrast to the distributed mixing described as an integrated gaussian used by Robbins [16]. Use of a localized mixing functional form is preferred in the present study since the experiment consisted of a single specie of animal all about the same size. Therefore these animals would be expected to feed at about the same depth in about the same manner at about the same rate, i.e., localized. Mixed specie assemblages and mixed size animals might more appropriately be described using the integrated form of Equation 9. Using a SYSTAT® nonlinear estimation program the parameters D_{bmax}, μ, and σ can be estimated for each of the *Yoldia* size classes. Figure 18 shows D_b as a function of depth for the larger sized *Yoldia*. Note that the data in Figure 18 can not only be approximated as a gaussian as we have done, but could also be reasonably well described by an integrated gaussian, exponential or even a linear value over a finite mixed zone.

Table 2. Biogenic Diffusion Coefficients, D_b, of Two Sizes of *Yoldia*. Values are slopes of the lines as shown for the larger size *Yoldia* in Figure 17 and noted in Equation 7.

Yoldia size (cm)	Peak	Peak Depth (cm)	D_b (cm²/yr)
	1	0.5	4.25
Large	2	2.4	1.12
(2.3-2.7 cm)	3	3.7	0.91
	4	5.6	0.05
	1	0.0	0.37
Small	2	1.3	0.05
(1.0-1.8 cm)	3	2.3	0.05

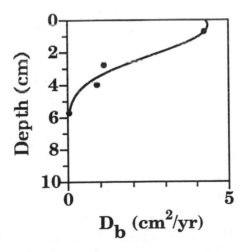

Figure 18. *Plots of the Slopes of the Lines in Figure 17 as a Function of Sediment Depth. A gaussian has been fit to the data based on the assumption that the majority of mixing activity occurs around an animal's life position. These data indicate that the majority of mixing by Yoldia occurs beneath the sediment-water interface at a depth of about 1 cm.*

This explains why numerous previous studies which have assumed a linear, exponential, gaussian or other functional forms of the biogenic diffusion coefficient and/or defined a mixed zone have been reasonably successful in calculating particle mixing. The approach employed here, however, is a particularly significant advance in bioturbation studies, because it not only allows the quantitative determination of the depth dependency of the mixing coefficient it demonstrates the exact form of that coefficient.

Reworking is also measured by the removal of labeled material from each layer. The tracer feeding rate is in turn related to the rate of particle feeding and velocity of peak burial. Losses of tracer from the feeding zone are accompanied by increases in other layers. Examination of the time series data (Figure 16) shows that the peak heights and areas decrease with time. This occurs because the animals not only cause diffusive mixing (peak broadening), but actually feed at depth in the sediment and eject the waste into the overlying water. After a clam ingests a portion of a tracer layer and ejects the material to the surface, the labeled peak no longer has as much tracer. This is represented as a decrease in peak area. Assuming that the clam does not selectively feed on or reject the tracer, the proportion of tracer that is removed from a depth interval is directly proportional to the concentration of the tracer in that depth interval. This pattern of feeding has been described as first-order feeding kinetics [15] and can be described as

$$\frac{dC}{dt} = -\gamma t \tag{10}$$

where these parameters have been previously described in Equation 3. Integrating Equation 10,

$$\ln C = \ln C_0 - \gamma t \tag{11}$$

where C_0 is the initial concentration of the tracer. The tracer concentration, C, of each peak was obtained from the area beneath it. Figure 19 shows that the peak areas decrease with time in accordance with this concept, and it also shows that the rate of decrease is less at depth, indicating that the clams are feeding faster closer to the sediment-water interface. Equation 11 indicates that a plot of $\ln C$ vs. t for each peak should be linear and the slope of the lines with depth yields the depth dependency of the feeding coefficient. Then, by plotting every γ versus the mean depth of each peak it is possible to obtain the feeding distribution profile (Figure 20). The functional form of the feeding distribution is unknown, although it seems reasonable to assume that it is gaussian - a subsurface deposit feeder can be expected to feed at a finite depth with a normal distribution about that depth:

$$\gamma(z) = \gamma_{max} \exp\left(-(z-\mu)^2 / 2\sigma^2\right) \tag{12}$$

Robbins [16] notes that feeding can be described as either localized (gaussian) or distributed (integrated gaussian). *Yoldia* is mostly a sedentary deposit feeder, removing sediment from around its fixed position just beneath the sediment-water interface and depositing sediment at the sediment surface. In these single specie experiments the animals were all about the same size, so use of a localized mixing functional form seems reasonable. Using a SYSTAT® nonlinear estimation program we estimated the parameters γ_{max}, μ, and σ for each of the *Yoldia* size classes. Figure 19 shows γ as a function of depth for the larger sized *Yoldia*. This approach is a particularly significant advance in bioturbation studies, because it permits the quantitative determination of the depth dependency of feeding.

Because conveyor belt species continuously deposit sediment onto the sediment-water interface, tracer layers within the zone of feeding will be buried at a rate defined by the feeding rate. This can be expressed as a vertical advection velocity which can readily be determined from a plot of the peak positions with time. This has not been shown here.

It is also possible to use Equation 3 to model the tracer profiles to obtain values for the diffusion coefficients and feeding rates for comparison with the experimentally-derived values. All other terms in the equation are known or have assumed values (Table 3). However, in order to compare modeled profiles with gamma scan data it is necessary to incorporate detector optics into the model since the observed data are subject to a broadening effect. Model profiles determined using Equation 3, a detector optics correction, and the parameters listed in Table 3 for the larger sized *Yoldia* are shown as the solid lines in Figure 16.

Generally speaking, the modeled activity versus depth profiles fit the experimental data very successfully. The only experimentally-derived parameters were the biogenic diffusion coefficients and the feeding rate coefficients. The experimentally-derived biogenic diffusion coefficients were determined from the time rate of spread in the tracer peak activities and the experimentally-derived feeding rate parameters were obtained from the time rate of change in peak areas. This means

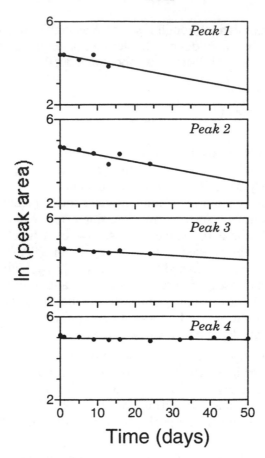

Figure 19. *Plots of ln (peak area) Versus Time for Each Peak
(Equation 11) for Selected Times in the Larger-sized
Yoldia Experiment. Slopes of the lines represent the rate
of decrease in area of an individual peak caused by feeding
on the tracer in that layer. Note that the slopes of the lines
are greatest beneath the surface indicating that Yoldia feeds
at depth in the sediment.*

that the model describes the major processes of sediment mixing in these
experiments, and that this experimental and mathematical approach can
be used to define the major mixing modes of individual species. In the
larger sized *Yoldia* cells the particle redistribution mainly occurs in the
depth range of 0-5 cm and is centered at a depth of about 2 cm. The
tracer activities in the top three layers decrease very quickly and
identifiable layers are destroyed within 50 days. Below the top 5 cm,
the activities in the tracer layers decrease much more slowly and could
be, by model prediction completely mixed only after 5 years. The long-
term model predictions are included to illustrate that one of the

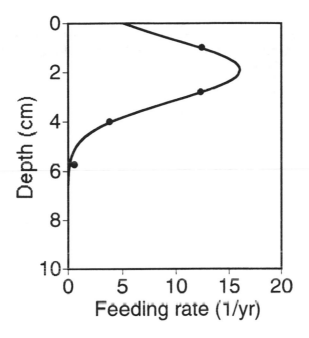

Figure 20. *Plots of the Slopes of the Lines in Figure 19 as a Function of Sediment Depth. A gaussian has been fit to the data based on the assumption that the majority of the animals feed around their life position. These data indicate that the majority of feeding by <u>Yoldia</u> occurs beneath the sediment-water interface at a depth of about 2 cm.*

major purposes of modeling is to make predictions. Obviously, after 5 or 10 years other processes would need to be considered, as well, and these model predictions can only be considered to illustrate some aspects of long-term mixing. In the case of no sediment accumulation from the water column the final activity profile after 10 years becomes a uniform plateau extending from the sediment-water interface to a depth of about 7.5 cm and then quickly declines to 0 at about 8 cm. In the small sized *Yoldia* cells the mixing of sediment is mostly centered within the top 1.5 cm of the sediment. Below this depth sediment redistribution is very slow and becomes significant only after about a year. The tracer activities below 1.5 cm and above 4 cm still decrease, although slowly, and are evened out after about 2 years. This probably means that the smaller *Yoldia* might occasionally mix sediment to greater depths, but not very often. Below 4 cm the peak shape is quite stable and only after about 5 years is significant change of peak shape apparent.

Table 3. Parameter Values Used in the Model Fits Shown in Figure 16.		
Sediment Properties		
ρ_s	(g/cm^3)	2.5
ω	(g/cm^2/year)	0.0
ν	(cm/year)	0.0
$\phi\,(0)$	---	0.75
$\phi\,(\infty)$	---	0.66
β	(cm^{-1})	0.016
Biogenic Coefficients		
$D_{b_{max}}$	(cm^2/year)	5.0
μ	(cm)	0.1
σ	(cm)	3.0
γ_{max}	(year^{-1})	6.5
μ	(cm)	2.0
σ	(cm)	2.0
η	---	1.0
Tracer Coefficients		
D_m	(cm^2/year)	157.0
λ	(year^{-1})	0.023
K	---	7.40×10^8
Computational Coefficients		
dt	(days)	0.5
t	(days)	60.0
dz	(cm)	0.2
z	(cm)	10.0

The modeled biodiffusion coefficients for the larger sized and smaller sized *Yoldia* are 1.67 (cm^2/yr/animal) and 0.05 (cm^2/yr/animal), respectively. These values are in excellent agreement with the results based on the rate of peak spreading (large *Yoldia*: 1.46 cm^2/yr/animal; small *Yoldia*: 0.049 cm^2/yr/animal). Such a good agreement is not a surprise because *Yoldia* is primarily an advector specie. Therefore, the tracer activity profile is controlled more by its advective feeding on sediment than its diffusional activities such as burrowing and changing position of living. However, the effects of animal size on the biogenic diffusion coefficient are significant: the larger sized *Yoldia* has a biogenic diffusion coefficient more than 30 times greater than that of the smaller animals. This is probably because the larger animals expend greater energy in sediment mixing that do the smaller ones. Second, the maximum diffusion depth (μ) and the diffusion spread (σ) are different. Larger size *Yoldia* have a smaller diffusion depth but greater spread (μ=0.35 cm; σ=2.0 cm) than the smaller animals (μ=0.1.4 cm; σ=1.0 cm). One possible explanation is that when the clams were added to the cells at the beginning of the experiment, they started to burrow on the sediment surface. Because of its larger size and higher energy, the larger sized *Yoldia* exhibited a larger diffusional spread (σ) than the small sized *Yoldia*. As the experiment continued the animals ceased burrowing activities and began conveyor-belt feeding. Thus, there was a change in the style of mixing from mostly diffusive mixing to advective particle transport. However, the effect of advective feeding by the smaller *Yoldia* on sediment redistribution is, to some extent, similar to that of their diffusional activities simply because of their small size. The difference, however, is that the maximum depth of mixing during advective feeding is deeper than the initial surface diffusion.

The difference between the modeled feeding rate and the feeding rate based on changes in peak area is larger for the larger sized *Yoldia* than for the smaller sized *Yoldia* (large *Yoldia*: modeled γ=2.67 sed frac/yr/animal; experimental γ=4.0 sed frac/yr/animal; small *Yoldia*: modeled γ=1.0 sed frac/yr/animal; experimental γ=1.07 sed frac/yr/animal). There are two reasons for this. First, the larger *Yoldia* generally have a higher feeding rate than the smaller ones, so the tracer layers within the feeding zone of the larger *Yoldia* are smeared and deformed faster than the rate of destruction by the smaller *Yoldia*. This causes the tracer layers to disappear and become hard to identify and follow during the experiment. As a result the measured data for the larger sized *Yoldia* are good only in the early stage of the experiment. Second, the number of animals used differed between the two sizes (3 animals in the larger size cells; 8 animals in the small size cells). Since the experimentally-measured activities are caused by the integrated effects of all animal activities, too large an animal size and/or too small a number of animals will both introduce random experimental error. Therefore, for the larger sized *Yoldia* the modeled feeding rate is more reliable than that determined from the changes in peak area. These results also indicate that the larger sized *Yoldia* have a higher feeding rate (γ=2.67), feed deeper (μ=2 cm), and feed over a bigger depth range (σ=2 cm) than the little ones (γ=1, μ=1 cm, σ=0.5 cm).

Another application of this type of model is that it may be used to simulate the effects of burial and mixing on the contaminant concentrations in layered sediments, such as those associated with capping of dredge disposal sediments. For example, it might be assumed that uncapped contaminated sediments will always retain higher concentrations of the contaminant in the surficial layer than contaminated sediments capped with a "clean" layer. This is not necessarily true, however, because recycling of a contaminated layer by conveyor-belt feeding and particle selectivity can result in higher concentrations in capped sediment under some conditions. Thus, depending on the nature and rates of the sediment redistribution processes, the kind and quantity of benthos involved, the thicknesses of the layers, and the time scales of interest, there may be situations when it is not desirable to cap the sediment. Figure 21 illustrates a case where a 10 cm layer of "clean" sediment overlies a 10 cm thick layer of contaminated sediment. These sediments are assumed to be mixed by tubificid oligochaetes only, which selectively feed on tracer-laden particles at a depth of 10 cm. These results indicate that rapid mixing into the contaminated zone causes recycling and ultimate destruction of a tracer concentration peak in a sediment layer, unless the thickness of the layer is greater than that of the feeding zone. Particle selective feeding causes a depletion of the tracer at the base of the zone of maximum feeding and an increase in its concentration above that depth. When coupled with advective recycling, particle selectivity causes the tracer to be preferentially retained within the mixed zone. Feeding at depths less than the thickness of the dredged disposal layer results in the burial of an unmixed lower portion of the layer. This has significant implications for possible regulations requiring capping of dredged sediment. For example, the mixing depth of marine fauna will vary with time since disturbance ([64] Figure 4), and although many marine species are confined to the upper 20 cm or so, some species such as callianassid shrimp have been found to burrow to depths as much as 1-3 meters.

Despite the relatively large and growing number of observations of natural tracers of mixing in sediment cores, it is not common that the benthic assemblages from which the cores are collected are assayed at the same time the cores are collected. When they are, the increased knowledge can often explain apparent anomalies in tracer distribution (e.g., [124,174,180,181,184]). Even less common are laboratory studies to show the particle mixing styles of dominant macrobenthos. Laboratory studies of mixing have usually consisted of the measurement

CONCENTRATION

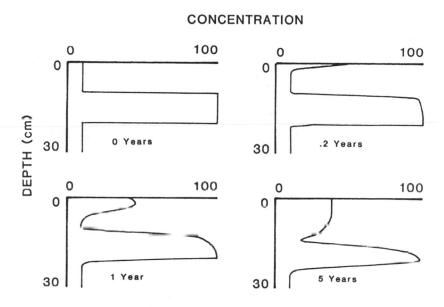

Figure 21.   *Example Simulation to Illustrate the Application of Bioturbation Modeling to Capping of Dredged Disposal Sediments. Model calculations show time evolution of a contaminant distribution. System consists of a 10 cm cap of "clean" sediment overlying a 10 cm contaminated layer which overlies "clean" background sediment. Bioturbating organisms are tubificid oligochaetes which selectively feed on the contaminated particles at a depth of 10 cm.*

of the volume or mass of fecal material ([19,145,170,185-190], and many others).

While these studies are important and useful for a number of purposes, they tell us little about the redistribution of mass within the zone of active mixing, which is required to predict the fate of contaminants introduced into the sediment. A second sort of experiment consists of the introduction of tracer particles into the sediment - usually at the sediment surface in the field or in laboratory microcosms - and the tracing of their redistribution by macrobenthos. A variety of marker horizons have been used, including colored particles [174,191-194], pollen grains [195], copepod eggs [193], exotic sedimentary minerals [19,172,196], rare earth elements [197], and radioactively labeled particles [162,198-200]. These studies, while they markedly improve our understanding of the way that infauna redistribute particulates, still suffer from some shortcomings. A problem with most of the particulate markers is one of sensitivity. Once particles reach the zone of maximum feeding/mixing the concentration of tracers becomes too low to follow the details of the mixing process. Another problem with all of these techniques is that experimental cells need to be destroyed for analysis, which decreases the reproducibility of results, and analysis of tracer concentration is often a tedious process. The work presented here demonstrates that the use of radionuclide tracer sediments can be used to not only follow the sediment to gain insight to the mixing process, but also to quantitatively describe the diffusive mixing and feeding processes.

The major problem utilizing any mathematical model is that many of the parameters (especially D_b and γ) are not known well and are certainly not known as a function of sediment depth. However, it was shown above that it is possible to quantify these bioturbation parameters by careful design of experiments so that the model coefficients and their functional forms can be determined independently. These experiments also reveal that *Yoldia* are significant conveyor-belt deposit feeders. The intensity and the range of sediment reworking are closely related to animal size: larger sized *Yoldia* mix sediments in the top 6 cm and feed on sediment at a rate of about 2.7 sediment fractions/animal/yr. at a feeding depth of about 2 cm while smaller *Yoldia* mix sediments in the top 3 cm at a feeding rate of about 1.0 sediment fractions/animal/yr.

2.7   Mathematical models of solute transport

The activities of benthos not only redistribute sediment particles, but as was discussed above, their actions can also influence the exchange of water and solutes across the sediment-water interface. Burrows act as channels for the direct communication between interstitial and overlying waters; fecal pellets deposited on the sediment surface increase the porosity of the uppermost sediment and therefore increase the diffusional exchange; and the injection of surface water for burrowing and feeding modifies the porosity and chemical conditions in the sediment and increases the exchange of water and solutes across the sediment-water interface. Permanent or semi-permanent tube structures or significant directed exchange of water cannot necessarily be described by simple one-dimensional models although models have been developed which employ an enhanced molecular diffusivity in the bioturbated zone [124]. This was successfully employed by McCall and Fisher [19] as enhanced diffusion through a tubificid pellet layer at the sediment surface and in Figure 8 for enhanced Cl⁻ transport by unionids. Other models, termed biopumping models [84] describe the exchange of solutes between well-mixed and distinct reservoirs of sediment and overlying water as an advection velocity [115-117]. Finally, Aller's [95] cylindrical burrow model attributes enhanced solute transport to shorter diffusional transport paths to/from irrigated burrows. This was demonstrated in Figures 12 and 14 for ammonium and manganese transport around chironomid burrows. Because of these different conceptual interpretations of solute transport in sediments it is necessary to devise better methods to quantify the transport processes and to test the competing models. The gamma scanning system shown in Figure 15 and described above for particle transport can also be used to track solute transport. Figure 22 shows some preliminary data from an experiment with *Yoldia limatula* in which solute transport is traced using ^{22}Na. The data indicate that *Yoldia* enhance solute exchange between sediments and overlying water on a time scale that is quite rapid (transport down the entire microcosm length of 15 cm was achieved within 15 days). This enhancement is most apparent at short times, but clearly persists during the entire experiment. Although these data have not been mathematically modeled yet, and *Yoldia* is not a benthic specie which irrigates a burrow, these results illustrate that this methodology of collecting closely spaced time and position data can be very useful for quantifying the exchange of solutes between sediments and water and for acquiring the kind and quality of data necessary to test different models of the solute exchange process.

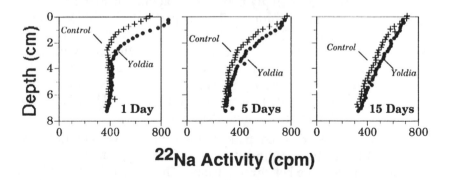

Figure 22.   Gamma Scan Data for ^{22}Na in the Yoldia Experiment. Enhanced solute exchange by the clams throughout the entire experiment is evident by the faster downward migration of ^{22}Na into the sediment in the presence of clams compared to the control.

3.   SUGGESTIONS FOR FUTURE RESEARCH

As we acquire a more detailed knowledge of processes that occur near the sediment-water interface we learn that the animal-sediment-water system is very complex. Some portions of the system have received a great deal of study and are reasonably well understood (for example, experimental methodologies to determine particle mixing by benthos) while other aspects of the system are essentially unstudied (for example, the relationship between bioturbation and metal sulfide reactions). Accordingly, progress in this system will require building on the knowledge base that has been established to address ever broader issues by establishing the linkages to apply theoretical and laboratory findings to actual field situations.

Additional laboratory bioturbation studies are still needed to integrate particle and solute transport into a single comprehensive model. In order to accomplish this, additional studies are needed to more fully understand selective feeding by benthos, especially particle size and food quality selectivity; to characterize mixing and solute transport by a wider range of benthos than the few organisms that have been studied; to characterize mixing and solute transport by mixed benthic assemblages, especially those found in natural field cores; and to determine the complex relationships between both particle mixing and solute transport and mineral precipitation/dissolution reactions, oxidation/reduction reactions, and microbially-mediated biogeochemical reactions in sediments. Despite the frequent use of solute transport modeling, we have a surprisingly poor understanding of solute transport in the presence of benthic animals. This is particularly obvious in the case of metals transport and control. Several mathematical methods are currently employed to describe solute transport as affected by benthos, but the physical interpretations of these models are significantly different. If these models are to receive a more universal application then additional work is needed to test and verify them. Then they need to be integrated with particle mixing models.

From a practical perspective the relationships between bioturbation and chemical fluxes from sediments and surface sediment contaminant concentrations need to be better established. There is sufficient evidence to indicate that bioturbation significantly affects the mass flux from sediments of many species, but the benthic community is dynamic so the chemical fluxes from sediments in the long-term are unknown. The implications of this in establishing chemical budgets is obvious. Similarly, it is reasonably well established that bioturbation can affect surface concentrations of particle-bound contaminants, but the application of this knowledge to pollutant bioaccumulation studies and to prediction of contaminant transport has not been established.

Perhaps the most striking finding is the paucity of knowledge relating bioturbation with the metal chemistry of sediments. The majority of studies of metals in sediments do not even address bioturbation, and most of those that do only incorporate a particle diffusion coefficient in a surface mixed layer or assume sulfate reduction only begins at a depth below some finite thickness of a bioturbated zone. There are only a handful of studies that actually examine the linkages between bioturbation and reactions involving metals in sediments (for example, Sundby and Silverberg [201]). This limited understanding of the linkages between bioturbation and metal chemistry in sediments is even more pronounced when metal sulfide chemistry is considered. Bioturbation is likely to cause major modifications of oxidation/reduction reactions in sediments and significantly affect the equilibria, kinetics, and distribution of metal sulfides. Clearly more work in this area is necessary. The implications of this in the establishment of sediment quality criteria based on metal sulfide chemistry are enormous and need to be carefully considered before any sediment quality criteria are enacted.

Acknowledgments: Portions of this work have been supported by different grants and contracts to the author and his coworkers. These include NSF Grant OCE-8005103, NOAA Grant NA80RAD00036, EPA Contract #R8057160, Ohio Sea Grant USURF 717649, Army Corps of Engineers Waterways Experimental Station Contract #DA CW39-84-M-3522, and EPA Contract #R-817278-01-0.

REFERENCES

1. McCall, P.L. and M.J.S. Tevesz, Eds. *Animal Sediment Relations: The Biogenic Alteration of Sediments* Plenum Press, (New York, 1982).

2. Fisher, J.B., "Effects of Macrobenthos on the Chemical Sediment Diagenesis of Freshwater Sediments," in *Animal-Sediment Relations*, P.L. McCall and M. Tevesz, Eds., pp. 177-218 Plenum Press, (New York, 1982).

3. Reynoldson, R.B., D.W. Schloesser and B.A. Manny, "Development of a Benthic Invertebrate Objective for Mesotrophic Great Lakes Waters," *J. Great Lakes Res.* 15:669-686 (1989).

4. McCall, P.L. and F.M. Soster, "Benthos Response to Disturbance in Western Lake Erie: Regional Faunal Surveys," *Can. J. Fish. Aquat. Sci.* 47:1996-2009 (1990).

5. Soster, F.M. and P.L. McCall, "Benthos Response to Disturbance in Western Lake Erie: Field Experiments," *Can. J. Fish. Aquat. Sci.* 47:1970-1985 (1990).

6. Robbins, J.A., P.L. McCall, J.B. Fisher and J.R. Krezoski, "Effect of Deposit-feeders on the Migration of Cs-137 in Lake Sediments," *Earth Planet. Sci. Lett.* 42:277-287 (1979).

7. Edwards, R.W. and H.L.J. Rolley, "Oxygen Consumption of River Muds," *J. Ecol.* 53:1-19 (1965).

8. Hargrave, B.T., "Stability in Structure and Function of the Mud-water Interface," *Verh. Int. Verein. Limnol.* 19:1073-1079 (1975).

9. Graneli, W., "The Influence of *Chironomus plumosus* Larvae on the Exchange of Dissolved Substances Between Sediment and Water," *Hydrobiologia* 66:149-159 (1979).

10. Graneli, W., "The Influence of *Chironomus plumosus* Larvae on the Oxygen Uptake of Sediment," *Arch. Hydrobiol.* 87:385-403 (1979).

11. Matisoff, G., J.B. Fisher and S. Matis, "Effects of Benthic Macroinvertebrates on the Exchange of Solutes Between Sediments and Freshwater," *Hydrobiologia* 122:19-33 (1985).

12. Fukuhara, J. and M. Sakamoto, "Enhancement of Inorganic Nitrogen and Phosphate Release from Lake Sediment by Tubificid Worms and Chironomid Larvae," *Oikos* 48:312-320 (1987).

13. McCall, P.L., M.J.S. Tevesz and S.F. Schwelgien, "Sediment Mixing by *Lampsilis radiata siliquoidea* (Mollusca) from Western Lake Erie," *J. Great Lakes Res.* 5:105-111 (1979).

14. Krezoski, J.R., J.A. Robbins and D.S. White, "Dual Radiotracer Measurement of Zoobenthos Mediated Solute and Particle Transport in Freshwater Sediments," *J. Geophys. Res.* 89:7937-7947 (1984).

15. Krezoski, J.R. and J.A. Robbins, "Vertical Distribution of Feeding and Particle-selective Transport of Cs-137 in Lake Sediments by Lumbriculid Oligochaetes," *J. Geophys. Res.* 90:11,999-12,006 (1985).

16. Robbins, J.A., "A Model for Particle Selective Transport of Tracers in Sediments with Conveyor-belt Deposit Feeders," *J. Geophys. Res.* 91:8542-8558 (1986).

17. McCall, P.L., M.J.S. Tevesz and G. Matisoff, "Effects of a Unionid Bivalve on the Physical, Chemical, and Microbial Properties of Cohesive Sediments from Lake Erie," *Amer. J. Sci.* 286:127-159 (1986).

18. Boyer, L.F., P.L. McCall, F. M. Soster and R.B. Whitlatch, "Deep Sediment Mixing by Burbot (*Lota lota*), Caribou Island Basin, Lake Superior, USA," *Ichnos* 1:91-95 (1990).

19. McCall, P.L. and J.B. Fisher, "Effects of Tubificid Oligochaetes on Physical and Chemical Properties of Lake Erie Sediments," in *Aquatic Oligochaete Biology*, R.O. Brinkhurst and D.G. Cook, Eds., pp. 253-317 Plenum Press, (New York, 1980).

20. Fisher, J.B. and G. Matisoff, "High Resolution Profiles of pH in Recent Sediments," *Hydrobiologia* 79:277-284 (1981).

21. Lyman, F.E., "Swimming and Burrowing Activities of Mayfly Nymphs of the Genus Hexigenia," *Ann. Entomol. Soc. Amer.* 26:250-256 (1943).

22. Walshe, B.M., "Feeding Mechanisms of Chironomous Larvae," *Nature* 160:474 (1947).

23. Johnson, M.S. and F. Munger, "Observations on Excessive Abundance of the Midge *Chironomus plumosus* at Lake Pepin," *Ecology* 11:110-126 (1930).

24. Hilsenoff, W.L., "Ecology and Population Dynamics of *Chironomus plumosus* (Diptera, Chironomidae) in Lake Winnebago, Wisconsin," *Ecology* 60:1183-1194 (1967).

25. Jonasson, P.M., "Ecology and Production of the Profundal Benthos in Relation to Phytoplankton in Lake Esrom," *Oikos Suppl.* 14:1-148 (1972).

26. Mackey, A.P., "Growth and Development of Larval Chironomidae," *Oikos* 28:270-275 (1977).

27. Cole, G.A., "Notes on the Vertical Distribution of Organisms in the Profundal Sediments of Douglas Lake, Michigan," *Amer. Midl. Nat.* 49:252-56 (1953).

28. Hilsenoff, W.L., "The Biology of *Chironomus plumosus* in Lake Winnebago, Wisconsin," *Ann. Entomol. Soc. Amer.* 59:465-473 (1966).

29. Ford, J.B., "The Vertical Distribution of Chironomidae in the Mud of a Stream," *Hydrobiologia* 19:262-272 (1967).

30. Walshe, B.M., "The Feeding Habit of Certain Chironomid Larvae Subfamily Tendipeninae," *Proc. Zool. Soc.* 121:63-79 (1951).

31. Mundie, J.H., "The Diurnal Activity of the Larger Invertebrates at the Surface of Lake la Ronge, Saskatchewan," *Can. J. Zool.* 37:945-956 (1959).

32. Ganapti, S.V., "The Role of the Bloodworm, *Chironomus plumosus*, in Accounting for the Presence of Phosphorous and Excessive Free Ammonia in the Filtrates from the Slow Sand Filters of the Madras Water Works," *Zool. Soc. India J.* 6:41-43 (1949).

33. Edwards, R.W., "The Effect of Larvae on *Chironomus riparius* Meigen on Redox Potentials of Settled Activated Sludge," *Ann. Appl. Biol.* 46:457-464 (1958).

34. Tessenow, U., "Experimentaluntersuchengen zur Kieselsaureruckfuhrung aus dem Schlamm der See durch Chironomidenlarven (Plumosus-Gruppe)," *Arch. Hydrobiol.* 60:497-504 (1964).

35. Weissenbach, H., "Untersuchungen zum Phophorhaushalt eines Hochgebirgsees (Vorder Finstertaler See, Kuhtai, Tirol) unter besonderer Berucksichtigung der Sedimente," *Ph.D. Dissertation*, Leopold Franzens-Universitat, Innsbruck (1974).

36. Gallepp, G.W., J.F. Kitchell and S.M. Bartell, "Phosphorous Release from Lake Sediments as Affected by Chironomids," *Verh. Int. Verein. Limnol.* 20:458-465 (1978).

37. Gallepp, G.W., "Chironomid Influence on Phosphorous Release in Sediment-Water Microcosms," *Ecology* 60:547-556 (1979).

38. Gardner, W.S., T.F. Nalepa, D.R. Slavens and G.A. Laird, "Patterns and Rates of Nitrogen Release by Benthic Chironomidae and Oligochaeta," *Can. J. Fish. Aquat. Sci.* 40:259-266 (1983).

39. Anderson, J.M., "Importance of the Denitrification Process for the Rate of Degradation of Organic Matter in Lake Sediments," in *Interactions Between Sediments and Freshwater*, H.L. Golterman, Ed., pp. 357-362 The Hague: Dr. W. Junk b.v. Publishers, (1977).

40. Lawrence, G.B., M.J. Mitchell and D.H. Landers, "Effects of the Burrowing Mayfly, *Hexagenia* on Nitrogen and Sulfur Fractions in Lake Sediment Microcosms," *Hydrobiologia* 87:273 283 (1982).

41. Gulubkov, S.M. and T.M.Tiunova, "Dependence of the Respiration Rate Upon Oxygen Concentration in Water for Some Rheophilous Mayfly Larvae (Ephemoeroptera)," *Aquat. Insects* 11:147-151 (1989).

42. Wiley, M.J. and S.L. Kohler, "Positioning Changes of Mayfly Nymphs Due to Behavioral Regulation of Oxygen Consumption," *Can. J. Zool.* 58:618-622 (1980).

43. Howmiller, R.P. and A.M. Beeton, "Biological Evaluation of Environmental Quality, Green Bay, Lake Michigan," *J. Wat. Pollut. Cont. Fed.* 43:123-133 (1971).

44. Schneider, J.C., F.H. Hooper and A.M. Beeton, "The Distribution of Benthic Fauna in Saginaw Bay, Lake Huron," in *Proc. 12th Conf. on Great Lakes Res.*, pp. 80-90 (Intern. Assoc. Great Lakes Res. 9, 1969).

45. Carr, J.F. and J.K. Hiltunen, "Changes in the Bottom Fauna of Western Lake Erie from 1930 to 1961," *Limnol. Oceanogr.* 10:551-569 (1965).

46. Gerould, S. and S.P. Gloss, "Mayfly-mediated Sorption of Toxicants into Sediments," *Environ. Toxicol. Chem.* 5:667-673 (1986).

47. Landrum, R.F. and R. Poore, "Toxicokinetics of Selected Xenobiotics in *Hexagenia limbata*," *J. Great Lakes Res.* 14:427-437 (1988).

48. Dukerschein, J.T., J.G. Wiener, R.G. Rada and M.T. Steingraeber, "Cadmium and Mercury in Emergent Mayflies (*Hexagenia bilineata*) from the Upper Mississippi River," *Arch. Environ. Toxicol.* 23:109-116 (1992).

49. Steingraeber, M.T., T.R. Schwartz, J.G. Wiener and J.A. Lebo, "Polychlorinated Biphenyl Congeners in Emergent Mayflies from the Upper Mississippi River," *Environ. Sci. Technol.* 28:707-714 (1994).

50. Thornley, S., "Macrozoobenthos of the Detroit and St. Clair Rivers with Companions to Neighboring Waters," *J. Great Lakes Res.* 11:290-296 (1985).

51. Schloesser, D.W., "Zonation of Mayfly Nymphs and Coddisfly Larvae in the St. Marys River," *J. Great Lakes Res.* 14:227-233 (1988).

52. Thornley, S. and Y. Hamdy, "An Assessment of the Bottom Fauna and Sediments of the Detroit River," *Ontario Ministry of Environment* (1984).

53. Hiltunen, J.K. and B.A. Manny, "Distribution and Abundance of Macrobenthos in the Detroit River and Lake St. Clair, 1977," *Administrative Report No. 82-2,* U.S. Fish Wild. Serv., Nat'l Fish. Res. Cent.-Great Lakes, Ann Arbor, Mich. (1982).

54. Mackie, G.L., "Biology of the Zebra Mussel (*Dreissena polymorpha*) and Observations of Mussel Colonization on Unionid Bivalves in Lake St. Clair of the Great Lakes," in *Zebra Mussels: Biology, Impacts and Control*, T.F. Nalepa and D.W. Schloesser, Eds., pp. 153-165 Lewis Publishers, (Chelsea, 1993).

55. Foster, T.D., "Observations on the Life History of a Fingernail Shell of the Genus *Sphaerium*," *J. Morphol.* 53:473-497 (1932).

56. Thomas, G.J., "Growth in One Species of Sphaerid Clam," *Nautilus* 79:47-58 (1965).

57. Mackie, G.L., S.U. Qadri and A.H. Clarke, "Intraspecific Variations of Growth, Birth Periods, and Longevity of *Musculium securis* (Bivalvia: Sphaeriidae) Near Ottawa, Canada," *Malacologia* 15:433-446 (1976).

58. Heard, W.H., "Reproduction of Fingernail Clams (*Sphaeriidae*: *Sphaerium* and *Musculium*)," *Malacologia* 16:421-455 (1977).

59. Eggleton, F.E., "Productivity of the Profundal Benthic Zone in Lake Michigan," *Pap. Mich. Acad. Sci.* 22:593-611 (1937).

60. Monakov, A.K., "Review of Studies on Feeding of Aquatic Invertebrates Conducted at the Institute of Biology of Inland Waters, Academy of Sciences, USSR," *J. Fish. Res. Board Can.* 29:363-383 (1972).

61. Meier-Brook, C., "Substrate Relations in Some *Pisidium* Species (*Eulamellibranchiata: Sphaeriidae*), *Malacologia* 9:121-125 (1969).

62. Mitropolsky, V.I., "Notes on the Life Cycle and Nutrition of *Sphaerium corneum L. (Mollusca, Lamellibranchia)*," *Trans. Inst. Biol. Inland Waters Acad. Sci. USSR* 12:125-128 (1966).

63. McCall, P.L., "Community Patterns and Adaptive Strategies of the Infaunal Benthos of Long Island Sound," *J. Mar. Res.* 38:221-266 (1977).

64. Pearson, T.H. and R. Rosenberg, "Macrobenthic Succession in Relation to Organic Enrichment and Pollution of the Marine Environment," *Oceanogr. Mar. Biol. Annu. Rev.* 16:229-311 (1978).

65. Rhoads, D.C., P.L. McCall and J.Y. Yingst, "Disturbance and Production on the Estuarine Seafloor," *Amer. Scientist*, 66:577-586 (1978).

66. Rhoads, D.C., J.Y. Yingst and W. Ullman, "Seafloor Stability in Central Long Island Sound. Part 1. Temporal Changes in Erodibility of Fine-grained Sediment," in *Estuarine Interactions*, M. Wiley, Ed., pp. 221-224 Academic Press, (New York, 1978).

67. Rhoads, D.C. and L.F. Boyer, "The Effects of Marine Benthos on Physical Properties of Sediments: A Successional Perspective," in *Animal-Sediment Relations*, P.L. McCall and M.J.S. Tevesz, Eds., pp. 3-52 Plenum Press, (New York, 1982).

68. Grassle, J.F. and J.P. Grassel, "Opportunistic Life Histories and Genetic Systems in Marine Benthic Polychaetes," *J. Mar. Res.* 32:253-284 (1974).

69. McCall, P.L., "Spatial-temporal Distributions of Long Island Sound Infauna: The Role of Bottom Disturbance in a Near Shore Marine Habitat," in *Estuarine Interactions*, M.L. Wiley, Ed., pp. 191-219 Academic Press, (New York, 1978).

70. Zajac, R.N. and R.B. Whitlach, "Responses of Estuarine Infauna to Disturbance. I. Spatial and Temporal Variation of Initial Recolonization," *Mar. Ecol. Prog. Ser.* 10:1-14.

71. Taghon, G.L. and R.R. Greene, "Utilization of Deposited and Suspended Particulate Matter by Benthic "Interface" Feeders," *Limnol. Oceanogr.* 37:1370-1391 (1992).

72. Sanders, H.L., "Oceanography of Long Island Sound, 1952-54. X. The Biology of Marine Bottom Communities," *Bull. Bingham. Oceanogr. Coll.* 15:345-414 (1956).

73. Sanders, H.L., "Benthic Studies in Buzzards Bay. I. Animal-sediment Relationships," *Limnol. Oceanogr.* 3:245-258 (1958).

74. Sanders, H.L., "Benthic Studies in Buzzards Bay. III. The Structure of the Soft-bottom Community," *Limnol. Oceanogr.* 5:138-153 (1960).

75. Rhoads, D.C. and D.K. Young, "The Influence of Deposit Feeding Organisms on Sediment Stability and Community Trophic Structure," *J. Mar. Res.* 28:150-178 (1970).

76. Aller, R.C., "The Effects of Macrobenthos on Chemical Properties of Marine Sediment and Overlying Water" in *Animal-Sediment Relations: The Biogenic Alteration of Sediments*, P.L. McCall and M.J.S. Tevesz, Eds., pp. 53-102 Plenum Press, (New York, 1982).

77. Matisoff, G. and J.A. Robbins, "A Model for Biological Mixing of Sediments," *J. Geol. Ed.* 35:144-149 (1987).

78. McCave, I.N., "Biological Pumping Upwards of the Coarse Fraction of Deep-sea Sediments," *J. Sed. Petrol.* 58:148-158 (1988).

79. Wheatcroft, R.A., "Experimental Tests for Particle Size-dependent Bioturbation in the Deep Ocean," *Limnol. Oceanogr.* 37:90-104 (1992).

80. Lopez, G.R. and J.S. Levinton, "Ecology of Deposit-feeding Animals in Marine Sediments," *Quat. Rev. Biol.* 62:235-260 (1987).

81. Yager, P.L., A.R.M. Nowell and P.A. Jumars, "Enhanced Deposition to Pits: A Local Food Source for Benthos," *J. Mar. Res.* 51:209-236 (1993).

82. Mayer, L.M., P.A. Jumars, G.L. Taghon, S.A. Macko and S. Trumbore, "Low-density Particles as Potential Nitrogenous Foods for Benthos," *J. Mar. Res.* 51:373-389 (1993).

83. Goldberg, E.D. and M. Koide, "Geochronological Studies of Deep Sea Sediments by the Ionium/Thorium Method," *Geochim. Cosmochim. Acta* 26:417-450 (1962).

84. Matisoff, G., "Mathematical Models of Bioturbation," in *Animal-Sediment Relations*, P.L. McCall and M.J.S. Tevesz, Eds., pp. 289-330 Plenum Press, (New York, 1982).

85. Van Cappellen, P., J.-F. Gaillard and C. Rabouille, "Biogeochemical Transformations in Sediments: Kinetic Models of Early Diagenesis," in *Interactions of C,N,P and S Biogeochemical Cycles and Global Change, vol. 14*, NATO ASI Series, R. Wollast, F.T. Mackenzie and L. Chou, Eds., pp. 401-445 Springer-Verlag, (Berlin, 1993).

86. Fukumori, E., E.R. Christensen and R.J. Klein, "A Model for ^{137}Cs and Other Tracers in Lake Sediments Considering Particle Size and the Inverse Solution," *Earth Planet. Sci. Lett.* 114:85-99 (1992).

87. Bruland, K.W., "Pb-210 Geochronology in the Coastal Environment," *Ph.D. Dissertation.* Univ. California, San Diego, CA (1974).

88. Turekian, K.K., J.K. Cochran and D.J. DeMaster, "Bioturbation in Deep Sea Deposits: Rates and Consequences," *Oceanus* 21:34-41 (1978).

89. Peng, T.-H., W.S. Broecker and W.H. Berger, "Rates of Benthic Mixing in Deep-sea Sediment as Determined by Radioactive Tracers," *Quat. Res.* 11:141-149 (1979).

90. Santschi, P.H., Y.-H. Li, J.J. Bell, R.M. Trier and K. Kawtaluk, "Pu in Coastal Marine Environments," *Earth Plan. Sci. Lett.* 51:248-265 (1980).

91. Guinasso, N.L. and D.R. Schink, "Quantitative Estimates of Biological Mixing Rates in Abyssal Sediments," *J. Geophys. Research* 80:3032-3043 (1975).

92. Benninger, L.K., R.C. Aller, J.K. Cochran and K.K. Turekian, "Effects of Biological Sediment Mixing on the [210]Pb Chronology and Trace Metal Distribution in a Long Island Sound Sediment Core," *Earth Planet. Sci. Lett.* 43:241-259 (1979).

93. Olsen, C.R., H.J. Simpson, T.-H. Peng, R.F. Bopp and R.M. Trier, "Sediment Mixing and Accumulation Rate Effects on Radionuclide Depth Profiles in Hudson Estuary Sediments," *J. Geophys. Res.* 86:11020-11028 (1981).

94. Schink, D.R. and N.L. Guinasso, "Processes Affecting Silica at the Abyssal Sediment-water Interface," in *Actes des Colloques du C.N.R.S.: Biogeochimie de la Matiere Organique a l'Interface Eau-Sediment Marin*, R. Daumas, Ed., pp. 81-92 Sta. Mar. Endoume, (Marseilles, 1982).

95. Aller, R.C., L.K. Benninger and J.K. Cochran, "Tracking Particle Associated Processes in Near Shore Environments by Use of ^{234}Th/^{238}U Disequilibrium," *Earth Planet. Sci. Lett.* 47:161-175 (1980).

96. Cochran, J.K. and R.C. Aller, "Particle Reworking in Sediments from the New York Bight Apex - Evidence from the Th-234 - U-238 Disequilibrium," *Estuarine Coastal Mar. Sci.* 9:739-747 (1980).

97. Demaster, D.J., C.A. Nittrouer, N.H. Cutshall, I.L. Larsen and E.P. Dion, "Short Lived Radionuclide Profiles and Inventories from Amazon Continental Shelf Sediments," *EOS* 61:1004 (1980).

98. Martin, W.R. and F.L. Sayles, "Seasonal Cycles of Particle and Solute Transport Processes in Near Shore Sediments: ^{222}Rn/^{226}Ra and ^{234}Th/^{238}U Disequilibrium at a Site in Buzzards Bay, MA," *Geochim. Cosmochim. Acta* 51:927-943 (1987).

99. Krishnaswami, S., L.K. Benninger, R.C. Aller and K.L. Van Damm, "Atmospherically-derived Radionuclides as Tracers of Sediment Mixing and Accumulation in Near-Shore Marine and Lake Sediments: Evidence from ^{7}Be, ^{210}Pb, And ^{239,240}Pu," *Earth Planet. Sci. Lett.* 47:307-318 (1980).

100. Cutshall, N.H., I.L. Larsen, C.R. Olsen, C.A. Nittrouer and D.J. DeMaster, "Columbia River Sediment in Quinault Canyon, Washington - Evidence from Artificial Radionuclides," *Mar. Geol.* 71:125-136 (1986).

101. Anderson, R.F., S.L. Schiff and R.H. Hesslein, "Determining Sediment Accumulation and Mixing Rates Using ^{210}Pb, ^{137}Cs, and Other Tracers: Problems Due to Postdepositional Mobility or Coring Artifacts," *Can. J. Fish. Aquat. Sci.* 44:231-250 (1987).

102. Santschi, P.H., S. Ballhalder, K. Farrenkothen, A. Luck, S. Zingg and M. Sturm, "Chernobyl Radionuclides in the Environment: Tracers for the Tight Coupling of Atmospheric, Terrestrial, and Aquatic Geochemical Processes," *Environ. Sci. Technol.* 22:510-516 (1988).

103. Santschi, P.H., S. Ballhalder, S. Zingg, A. Luck and K. Farrenkothen, "The Self-cleaning Capacity of Surface Waters After Radioactive Fallout. Evidence from European Waters After Chernobyl, 1986-1988," *Environ. Sci. Technol.* 24:519-526 (1990).

104. Kansanen, P.H., T. Jaakkola, S. Kulmala and R. Suutarinen, "Sedimentation and Distribution of Gamma-Emitting Radionuclides in Bottom Sediments of Southern Lake Paijanne, Finland, After the Chernobyl Accident," *Hydrobiologia* 222:121-140 (1991).

105. Schuler, C., W. Wieland, P.H. Santschi, M. Sturm, A. Luck, S. Bollhalder, J. Beer, G. Bonani, H.J. Hofmann, M. Suter and W. Wolfli, "A Multitracer Study of Radionuclides in Lake Zurich, Switzerland: 1. Comparison of Atmospheric and Sedimentary Fluxes of ^{7}Be, ^{10}Be, ^{210}Pb, ^{210}Po, and ^{137}Cs," *J. Geophys. Res.* 96:17051-17065 (1991).

106. Wieland, E., P.H. Santschi and J. Beer, "A Multitracer Study of Radionuclides in Lake Zurich, Switzerland 2. Residence Times, Removal Processes, and Sediment Focusing," *J. Geophys. Res.* 96:17067-17080 (1991).

107. Robbins, J.A., J. Lindner, W. Pfeiffer, J. Kleiner, H.H. Stabel and P. Frenzel, "Epilimnetic Scavenging of Chernobyl Radionuclides in Lake Constance," *Geochim. Cosmochim. Acta* 56:2339-2361 (1992).

108. Wieland, E., P.H. Santschi, P. Hohener and M. Sturm, "Scavenging of Chernobyl ^{137}Cs and Natural ^{210}Pb in Lake Sempach, Switzerland," *Geochim. Cosmochim. Acta* 57:2959-2979 (1993).

109. Aller, R.C., "The Influence of Macrobenthos on Chemical Diagenesis of Marine Sediments," *Ph.D. Dissertation.* Yale University, New Haven, CT (1977).

110. Allanson, B.R., D. Skinner and J. Imberger, "Flow in Prawn Burrows," *Estuar. Coast. Shelf Sci.* 35:253-266 (1992).

111. Murphy, R.C. and J.N. Kremer, "Benthic Community Metabolism and the Role of Deposit-feeding Callianassid Shrimp," *J. Mar. Res.* 50:321-340 (1992).

112. Kristensen, E., R.C. Aller and J.Y. Aller, "Oxic and Anoxic Decomposition of Tubes from the Burrowing Sea Anemone *Ceriantheopsis americanus*: Implications for Bulk Sediment Carbon and Nitrogen Balance," *J. Mar. Res.* 49:589-617 (1991b).

113. Fisher, J.B., W. Lick, P.L. McCall and J.A. Robbins, "Vertical Mixing of Lake Sediments by Tubificid Oligochaetes," *J. Geophys. Res.* 85:3997-4006 (1980).

114. Kristensen, E., M.H. Jensen and R.C. Aller, "Direct Measurement of Dissolved Inorganic Nitrogen Exchange and Denitrification in Individual Polychaete (*Nereis virens*) Burrows," *J. Mar. Res.* 49:355-377 (1991).

115. Grundmanis, V. and J.W. Murray, "Nitrification and Denitrification in Marine Sediments from Puget Sound," *Limnol. Oceanogr.* 22:804-813 (1977).

116. Hammond, D.E. and C. Fuller, "The Use of Radon-222 as a Tracer in San Francisco Bay," in *San Francisco Bay: The Urbanized Estuary*, T.J. Comomos, Ed., pp. 213-230 Amer. Assoc. Adv. Sci., (San Francisco, 1979).

117. McCaffrey, R.J., A.C. Myers, E. Davey, G. Morrison, M. Bender, N. Luedtke, E. Cullen, B. Froelich and G. Klinkhammer, "The Relation Between Pore Water Chemistry and Benthic Fluxes of Nutrients and Manganese in Narragansett Bay, Rhode Island," *Limnol. Oceanogr.* 25:31-44 (1980).

118. Berner, R.A, *Early Diagenesis: A Theoretical Approach* Princeton University Press, (Princeton, 1980).

119. Hammond, D.E., H.J. Simpson and G. Mathieu, "Methane and Radon-222 as Tracers for Mechanisms of Exchange Across the Sediment-water Interface in the Hudson River Estuary," in *Marine Chemistry in the Coastal Environment*, T.M. Church, Ed., pp. 119-132 Amer. Chem. Soc. Symp. Ser. 18, Amer. Chem. Soc., (Washington, DC 1975).

120. Schink, D.R., N.L. Guinasso, Jr. and K.A. Fanning, "Processes Affecting the Concentration of Silica at the Sediment - Water Interface of the Atlantic Ocean," *J. Geophys. Res.* 80:3013-3031 (1975).

121. Goldhaber, M.B., R.C. Aller, J.K. Cochran, J.K. Rosenfeld, C.S. Martens and R.A. Berner, "Sulfate Reduction, Diffusion and Bioturbation in Long Island Sound Sediments: Report of the Foam Group," *Amer. J. Sci.* 277:193-237 (1977).

122. Schink, D.R. and N.L. Guinasso, "Modeling the Influence of Bioturbation and Other Processes on Calcium Carbonate Dissolution at the Sea Floor," in *The Fate of Fossil Fuel CO2 in the Oceans*, N.R. Andersen and A. Malahoff, Eds., pp. 375-398 Plenum Press, (New York, 1977).

123. Schink, D.R. and N.L. Guinasso, "Redistribution of Dissolved and Adsorbed Materials in Abyssal Marine Sediments Undergoing Biological Stirring," *Amer. J. Sci.* 278:687-702 (1978).

124. Aller, R.C., and J.Y. Yingst, "Biogeochemistry of Tube-dwellings: A Study of the Sedentary Polychaete *Amphitrite ornata* (Leidy)," *J. Mar. Res.* 36:201-254 (1978).

125. Cochran, J.K. and S. Krishnaswami, "Radium, Thorium, Uranium, and ^{210}Pb in Deep-sea Sediments and Sediment Pore Waters from the North Equatorial Pacific," *Amer. J. Sci.* 280:849-889 (1980).

126. Aller, R.C., "Quantifying Solute Distributions in the Bioturbated Zone of Marine Sediments by Defining an Average Microenvironment," *Geochem. Cosmochim. Acta* 44:1955-1965 (1980).

127. Aller, R.C., "Diagenetic Processes Near the Sediment-water Interface of Long Island Sound. I. Decomposition and Nutrient Element Geochemistry (S, N, P), in Estuarine Physics and Chemistry: Studies in Long Island Sound," in *Advances in Geophysics, vol. 22*, B. Saltzman, Ed., pp. 237-350 Academic Press, (New York, 1980).

128. Aller, R.C., "The Importance of the Diffusive Permeability of Animal Burrow Linings in Determining Marine Sediment Chemistry," *J. Mar. Res.* 41:299-322 (1983).

129. Christensen, J.P., A.H. Devol and W.M. Smethie, Jr., "Biological Enhancement of Solute Exchange Between Sediments and Bottom Water on the Washington Continental Shelf," *Cont. Shelf. Res.* 3:9-23 (1984).

130. Emerson, S., R. Jahnke, and D. Heggie. "Sediment-Water Exchange in Shallow Water Estuarine Sediments," *J. Mar. Res.* 42:709-730 (1984).

131. Boudreau, B.P., "On the Equivalence of Nonlocal and Radial-diffusion Models for Porewater Irrigation," *J. Mar. Res.* 42:731-735 (1984).

132. Martin, W.R. and G.T. Banta, "The Measurement of Sediment Irrigation Rates: A Comparison of the Br^- Tracer and $^{222}Rn/^{226}Ra$ Disequilibrium Techniques," *J. Mar. Res.* 50:125-154 (1992).

133. Clavero, V., F.X. Niell, and J.A. Fernandez, "Effects of *Nereis diversicolor* O.F. Muller Abundance on the Dissolved Phosphate Exchange Between Sediment and Overlying Water in Palmones River Estuary (S. Spain)," *Estuar. Coast. Shelf Sci.* 33:193-202 (1991).

134. Clavero, V., J.A. Fernandez and F.X. Niell, "Bioturbation by *Nereis* Sp. and Its Effects on the Phosphate Flux Across the Sediment-water Interface in the Palmones River Estuary," *Hydrobiologia* 235/236:387-392 (1992).

135. Henriksen, K., M.B. Rasmussen and A. Jensen, "Effect of Bioturbation on Microbial Nitrogen Transformations in the Sediment and Fluxes of Ammonium and Nitrate to the Overlying Water," *Envir. Biogeochem.* 35:193-205 (1983).

136. McCall, P.L. and M.J.S. Tevesz, "The Effects of Benthos on Physical Properties of Freshwater Sediments," in *Animal Sediment Relations: The Biogenic Alteration of Sediments*, P.L. McCall and M.J.S. Tevesz, Eds., pp. 105-176 Plenum Press, (New York, 1982).

137. Mortimer, C.H., "The Exchange of Dissolved Substances Between Mud and Water in Lakes," *J. Ecol.* 29:363-383 (1941).

138. Kikuchi, E. and Y. Kurihara, "In Vitro Studies on the Effects of Tubificids on the Biological, Chemical, and Physical Characteristics of Submerged Rice Field Soil and Overlying Water," *Oikos* 29:348-356 (1977).

139. Nriagu, J.O. and C.I. Dell, "Diagenetic Formation of Iron Phosphates in Recent Lake Sediments," *Amer. Mineral.* 56:1055-1061 (1974).

140. Chatarpaul, L., J.B. Robinson and N.K. Kaushik, "Role of Tubificid Worms in Nitrogen Transformations in Stream Sediment," *J. Fish. Res. Board Can.* 36:673-678 (1979).

141. Chatarpaul, L., J.B. Robinson and N.K. Kaushik, "Effects of Tubificid Worms on Denitrification and Nitrification in Stream Sediment," *Can. J. Fish. Aquat. Sci.* 37:656-663 (1980).

142. Quigley, M.A., "Silica Regeneration Processes in Near Shore Southern Lake Michigan," *Abstracts. 24th Conf. Great Lakes Res., Columbus, OH*:46 (1981).

143. Schelske, C.L., E.F. Stoermer, D.J. Conley, J.A. Robbins and R.M. Glover, "Early Eutrophication in the Lower Great Lakes: New Evidence from Biogenic Silica in Sediments," *Science* 222:320-322 (1983).

144. Rhoads, D.C. "Organism-Sediment Relations on the Muddy Sea Floor," *Oceanogr. Mar. Biol. Ann. Rev.* 12:263-300 (1974).

145. Thayer, C.W., "Sediment-mediated Biological Disturbance and the Evolution of Marine Benthos," in *Biotic Interactions in Recent and Fossil Benthic Communities*, M.J.S. Tevesz and P.L. McCall, Eds., pp. 480-625 Plenum Press, (New York, 1983).

146. Dobbs, F.C. and R.B. Whitlach, "Aspects of Deposit-feeding by the Polychaete *Clymenella torquata*," *Ophelia* 21:159-166 (1982).

147. Bender, K. and W.R. Davis, "The Effect of *Yoldia limatula* on Bioturbation," *Ophelia* 23:91-100 (1984).

148. Richardson, M.D., K.B. Briggs and D.K. Young, "Effects of Biological Activity by Abyssal Benthic Macroinvertebrates on a Sedimentary Structure in the Venezuela Basin," *Mar. Geol.* 68:243-267 (1985).

149. Hines, A.H. and K.L. Comtois, "Vertical Distribution of Infauna in Sediments of a Subestuary of Central Chesapeake Bay," *Estuaries* 8:296-304 (1985).

150. Tunberg, B., "Studies on the Population Ecology of *Upogebia deltaura* (Leach)(Crustacea, Thalassinidea)," *Estuar. Coastal Shelf Sci.* 22:753-765 (1986).

151. Mauviel, A., S.K. Juniper and M. Sibuet, "Discovery of an Enteropneust Associated with a Mound-burrows Trace in the Deep Sea: Ecological and Geochemical Implications," *Deep-Sea Res.* 34:329-335 (1987).

152. Takeda, S. and Y. Kurihara, "The Effects of Burrowing of *Helice Tridens* (De Haan) on the Soil of a Salt-marsh Habitat," *J. Exp. Mar. Biol. Ecol.* 113:79-89 (1987).

153. Nelson, C.H., K.R. Johnson and J.H. Barber, "Gray Whale and Walrus Feeding Excavation on the Bering Shelf, Alaska," *J. Sed. Petrol.* 57:419-430 (1987).

154. Schaffner, L.C., R.J. Diaz, C.R. Olsen and I.L. Larsen, "Faunal Characteristics and Sediment Accumulation Processes in the James River Estuary, Virginia," *Estuar. Coastal, Shelf Sci.* 25:211-226 (1987).

155. Rice, D.L. and D.C. Rhoads, "Early Diagenesis of Organic Matter and the Nutritional Value of Sediment," in *Ecology of Marine Deposit Feeders*, pp. 59-97 Springer-Verlag, (New York, 1988).

156. Boudreau, B.P., "Mathematics of Tracer Mixing in Sediments: I. Spatially Dependent, Diffusive Mixing," *Amer. J. Sci.* 286:161-198 (1986).

157. Boudreau, B.P., "Mathematics of Tracer Mixing in Sediments: II. Nonlocal Mixing and Biological Conveyor-belt Phenomena," *Amer. J. Sci.* 286:199-238 (1986).

158. Boudreau, B.P. and D.M. Imboden, "Mathematics of Tracer Mixing in Sediments: III. The Theory of Nonlocal Mixing within Sediments," *Amer. J. Sci.* 287:693-719 (1987).

159. Keilty, T.J., D.S. White and P.F. Landrum, "Sublethal Responses to Endrin in Sediment by *Stylodrilus heringianus* (Lumbriculidae) as Measured by a [137]Cesium Marker Layer Technique," *Aquatic Toxicol.* 13:251-270 (1988).

160. Keilty, T.J., D.S. White and P.F. Landrum, "Sublethal Responses to Endrin in Sediment by *Limnodrilus hoffmeisteri* (Tubificidae), and in Mixed-culture with *Stylodrilus heringianus* (Lumbriculidae)," *Arch. Environ. Contam. Toxicol.* 17:95-101 (1988).

161. Adler, D.M., "Tracer Studies in Marine Microcosms: Transport Processes Near the Sediment-water Interface," *Ph.D. Dissertation.* Columbia University, New York, NY (1982).

162. Santschi, P.H., U.P. Nyffeler, P. Ohara, M. Bucholtz and W.S. Broecker, "Radiotracer Uptake on the Sea Floor: Results from the MANOP Chamber Deployments in the Eastern Pacific," *Deep-Sea Res.* 29:953-965 (1984).

163. Nyffeler, U.P., P.H. Santschi and Y. Li, "The Relevance of Scavenging Kinetics to Modeling of Sediment-water Interactions in Natural Waters," *Limnol. Oceanogr.* 31:277-292 (1986).

164. McCall, P.L. and J. A. Robbins, Laboratory Studies of Bioturbation Using Radiotracers. Abstracts ASLO 92, (Santa Fe, New Mexico 1992).

165. McCall, P.L., G. Matisoff, X. Wang and J.A. Robbins, "Sediment Mixing by the Marine Clam, *Yoldia limatula*," *J. Mar. Res.* (in preparation).

166. McMurty, G.M., R.C. Schneider, P.L. Colin, R.W. Buddemeir and T.M. Suchanek, "Vertical Distribution of Fallout Radionuclides in Enewetak Lagoon Sediments: Effects of Burial and Bioturbation on the Radionuclide Inventory," *Bull. Mar. Sci.* 38:35-55 (1986).

167. Robbins, J.A. and D.N. Edgington, "Determination of Recent Sedimentation Rates in Lake Michigan Using Pb-210 and Cs-137," *Geochim. Cosmochim. Acta* 39:285-304 (1975).

168. Officer, C.B. and D.R. Lynch, "Interpretation Procedures for the Determination of Sediment Parameters from Time Dependent Flux Inputs," *Earth Planet. Sci. Lett.* 61:55-62 (1982).

169. Officer, C.B. and D.R. Lynch, "Bioturbation, Sedimentation and Sediment-water Exchanges," *Estuar. Coastal Shelf Sci.* 28:1-12 (1989).

170. Rhoads, D.C., "Rates of Sediment Re-working by *Yoldia limatula* in Buzzard's Bay, Massachusetts and Long Island Sound," *J. Sed. Petrol.* 33:723-727 (1963).

171. Young, D.K. and D.C. Rhoads, "Animal-sediment Relations in Cape Cod Bay, Massachusetts. I. A Transect Study," *Mar. Biol.* 11:242-254 (1971).

172. Rice, D.L., "Early Diagenesis in Bioadvective Sediments: Relationships Between the Diagenesis of Beryllium-7, Sediment Reworking Rates, and the Abundance of Conveyor-belt Deposit-feeders," *J. Mar. Res.* 44:149-184 (1986).

173. Suchanek, T.H., "Control of Seagrass Communities and Sediment Redistribution by Callianassa (Crustacea: Thalassinidea) Bioturbation," *J. Mar. Res.* 41:281-298 (1983).

174. Suchanek, T.H. and P.L. Colin, "Rates and Effects of Bioturbation by Invertebrates and Fishes at Eniwetok and Bikini Atolls," *Bull. Mar. Sci.* 38:25-34 (1986).

175. Riddle, M.J., "Cyclone and Bioturbation Effects on Sediments from Coral Reef Lagoons," *Estuar. Coastal Shelf Sci.* 27:687-695 (1988).

176. Benninger, L.K. and S. Krishnaswami, "Sedimentary Processes in the Inner New York Bight: Evidence from Excess ^{210}-Pb and ^{239,240}-Pu," *Earth. Planet. Sci. Lett.* 53:158-71 (1981).

177. Carpenter, R., M.L. Peterson and Bennett, "^{210}-Pb Derived Sediment Accumulation and Mixing Rates for the Greater Puget Sound Region," *Mar. Geol.* 64:291-312 (1985).

178. Smith, J.N. and C.T. Schafer, "Bioturbation Processes in Continental Slope and Rise Sediments Delineated by Pb-210, Microfossil and Textural Indicators," *J. Mar. Res.* 42:1117-1145 (1984).

179. Li, W.Q., N.L. Guinasso, K.H. Cole, M.D. Richardson, J.W. Johnson and D.R. Schink, "Radionuclides as Indicators of Sedimentary Processes in Abyssal Caribbean Sediments," *Mar. Geol.* 68:187-204 (1985).

180. Smith, J.H., B.P. Boudreau and V. Noshkin, "Plutonium and ^{210}-Pb Distributions in Northeast Atlantic Sediments: Subsurface Anomalies Caused by Non-local Mixing," *Earth Planet. Sci. Lett.* 81:15-28 (1986).

181. Swinbanks, D.D. and Y. Shirayama, "A Model of the Effects of an Infaunal Xenophyophore on ^{210}-Pb Distribution in Deep-sea Sediment," *La mer* 24:69-74 (1986).

182. Myers, A.C., "Sediment Processing in a Marine Subtidal Sandy Bottom Community. I. Physical Aspects," *J. Mar. Res.* 35:609-632 (1977).

183. Duncan, P.B., "Burrow Structure and Burrowing Activity of the Funnel-feeding Enteropneust *Balanoglossus aurantiacus* in Bogue Sound, North Carolina, USA," *Mar. Ecol.* 8:75-95 (1987).

184. Aller, J.Y. and R.C. Aller, "Evidence for Localized Enhancement of Biological Activity Associated with Tube and Burrow Structures in Deep-sea Sediments at the HEBBLE Site, Western North Atlantic," *Deep-Sea Res.* 33:755-790 (1986).

185. Appleby, A.G. and R.O. Brinkhurst, "Defecation Rate of Three Tubificid Oligochaetes Found in the Sediment of Toronto Harbour, Ontario," *J. Fish. Res. Bd. Canada* 27:1971-1982 (1970).

186. Hylleberg, J. "Selective Feeding by *Abarenicola pacifica* with Selective Notes on *Abarenicola vagabunda* and a Concept of Gardening in Lugworms," *Ophelia* 14:113-137 (1975).

187. Cammen, L.M., "Ingestion Rate: An Empirical Model for Aquatic Deposit Feeders and Detritovores," *Oecologia* 44:303-310 (1980).

188. Whitlach, R.B. and J.R. Weinberg, "Factors Influencing Particle Selection and Feeding Rate in the Polychaete *Cystenides (Pectinaria) gouldii*," *Mar. Biol.* 71:33-40 (1982).

189. Lopez, G.R. and I.J. Cheng, "Synoptic Measurements of Ingestion Rate, Ingestion Selectivity, and Absorption Efficiency of Natural Foods in the Deposit-feeding Molluscs *Nucula annulata* (Bivalvia) and *Hydrobia totteni* (Gastropoda)," *Mar. Ecol. Prog. Ser.* 11:55-62 (1983).

190. Rice, D.L., T.S. Bianchi and E.H. Roper, "Experimental Studies of Sediment Reworking and Growth of *Scoloplos spp.* (Orbiniidae: Polychaete)," *Mar. Ecol. Prog. Ser.* 30:9-19 (1986).

191. Darwin, C., "On the Formation of Vegetable Mould," *Trans. Geol. Soc.* London, 5:505 (1837).

192. Haven, D.S. and R. Morales-Alamo, "Use of Fluorescent Particles to Trace Oyster Biodeposits in Marine Sediments," *J. Cons. Perm. Int. Explor. Mar.* 30:237-269 (1966).

193. Marcus, N.H. and J. Schmidt-Gengenbach, "Recruitment of Individuals into the Plankton: The Importance of Bioturbation," *Limnol. Oceanogr.* 3:206-210 (1986).

194. Mahaut, M.L. and G. Graf, "Luminophore Tracer Technique for Bioturbation Studies," *Oceonol. Acta* 3:323-328 (1987).

195. Davis, R.B., "Stratigraphic Effects of Tubificids in Profundal Lake Sediments," *Limnol. Oceanogr.* 19:466-488 (1974).

196. Aller, R.C., and R.E. Dodge, "Animal-sediment Relations in a Tropical Lagoon Discovery Bay, Jamaica," *J. Mar. Res.* 32:209-232 (1974).

197. Krezoski, J., "Sediment Reworking and Transport in Eastern Lake Superior: *In Situ* Rare Earth Element Tracer Studies," *J. Great Lakes Res.* 15:26-33 (1989).

198. Sorokin, J.I., "Carbon-14 Method in the Study of the Nutrition of Aquatic Animals," *Int. Rev. Gesamten Hydrobiol.* 51:209-224 (1966).

199. Amiard-Triquet, C., "Etude Experimentale de la Contamination par le Cerium 144 et le fer 59 d'un Sediment a *Arenicola marina* L.(Annelide Polychete)," *Cxah. Biol. Mar.* 15:483-494 (1974).

200. Haven, D.S., R. Morales-Alamo and J.N. Kraeuter, "Sediment Mixing by Invertebrates as Shown by [85]-Kr," *Special Scientific Report No. 109*, p. 29 Virginia Inst. Marine Sciences (1981).

201. Sundby, B. and N. Silverberg, "Manganese Fluxes in the Benthic Boundary Layer," *Limnol. Oceanogr.* 30:372-381 (1985).

10

EPA'S CONTAMINATED SEDIMENT MANAGEMENT STRATEGY

Thomas M. Armitage
Office of Science and Technology
U.S. Environmental Protection Agency
Washington, DC 20460

1. INTRODUCTION

In the 1980s EPA documented the extent and severity of contaminated sediment problems at sites throughout the U.S. Concerned with the mounting evidence of ecological and human health effects, EPA's Office of Water organized a Sediment Steering Committee to manage contaminated sediments. EPA has prepared a comprehensive contaminated sediment management strategy to coordinate and focus EPA's resources on contaminated sediment problems [1]. The strategy is based upon three major principles: 1) In-place sediment should be protected from contamination to ensure that the beneficial uses of the nation's surface waters are maintained for future generations; 2) Protection of inplace sediment should be achieved through pollution prevention and source controls; 3) Natural recovery is the preferred remedial technique. In-place sediment remediation will be limited to high risk sites where natural recovery will not occur in an acceptable time period, and where the cleanup process will not cause greater problems than leaving the site alone. EPA's contaminated sediment management strategy includes several component elements: assessment, prevention, remediation, dredged material management, research, and outreach.

In surveys conducted in 1985 and 1987 [2,3], the Office of Water (OW) of EPA first began to document the extent and severity of sediment contamination. Most of the information in the surveys described areas in the Northeast, along the Coast of the Atlantic Ocean and Gulf of Mexico, and in the Great Lakes region. The surveys found that heavy metals and metalloids (e.g., arsenic), polychlorinated biphenyls, pesticides, and polycyclic aromatic hydrocarbons are the most frequently reported contaminants in sediments. Significant ecological impacts were often reported at contaminated sediment sites, including impairment of reproductive capacity, and impacts to the structure and health of benthic and other aquatic communities. Potential human health impacts were noted at a number of sites where fish consumption advisories or bans were issued [3]. In 1989, a study by the National Academy of Sciences entitled Contaminated Marine Sediments Assessment and Remediation [4], also identified the potential for far-reaching health and ecological effects from contaminated sediments.

Many potential sources of contaminants to sediments are identified in the reports cited above. These sources include: municipal sewage treatment plants; combined sewer overflows; stormwater discharges from municipal and industrial facilities; direct industrial discharges of process waste; runoff and leachate from hazardous and solid waste sites; agricultural runoff; runoff from mining operations; runoff from industrial manufacturing and storage sites; and atmospheric deposition of contaminants.

Many of the sediment data used in the EPA studies were collected prior to regular analysis for such parameters as grain size, total organic carbon, or acid volatile sulfides. Such data are needed to determine bioavailability of sediment contaminants.

Rarely is such information available for historical sediment data. In order to develop a better understanding of the extent and severity of sediment contamination in the United States, better data on sediment quality, as well as direct measurements of chemical concentrations in edible fish tissue and toxicity bioassays, are needed. Large quantities of both published and unpublished data on sediment quality have not been placed in accessible or usable form, and many locations in the country have not been adequately sampled. However, as discussed below, EPA is currently compiling available sediment quality data in a National Sediment Inventory. Several recent national and regional sediment monitoring programs, including EPA's Environmental Monitoring and Assessment Program, are currently collecting data on physical and chemical characteristics of sediments, parameters describing bioavailability of contaminants, toxicity bioassays, contaminant residues in aquatic organism tissues, and biological community structures.

As noted above, additional data are required to develop a better understanding of sediment contamination problems in the United States. However, it is evident from the best data currently available that sediments in many waterbodies across the country are contaminated to levels that harm benthic and aquatic communities and that may contribute to increased cancer and noncancer diseases for consumers of contaminated fish and shellfish. The effects of sediment contamination have been documented sufficiently to confirm their importance as a significant and widespread national problem. To further define the national extent of contamination, EPA, under the authority of Section 503 of the Water Resources Development Act of 1992, has developed a national inventory of contaminated sediment sites. A report to Congress on this inventory will be completed in 1995.

In 1990 the Environmental Protection Agency's Administrator became concerned with mounting evidence of environmental problems linked to sediment contamination. The Administrator formed an Agency-wide Sediment Steering Committee to address the problem of contaminated sediment on a national scale. The committee, chaired by the Assistant Administrator of EPA's Office of Water, is composed of senior managers from all program offices with the authority to address contaminated sediments, as well as a representative from each of EPA's 10 Regional offices. The Sediment Steering Committee prepared an Agency-wide Contaminated Sediment Management Strategy to coordinate and focus the Agency's resources on contaminated sediment problems. The purpose of the Strategy is to present a cross-program policy framework in which EPA can promote consideration and reduction of ecological and human health risks posed by sediment contamination; and to describe specific actions EPA can take to bring about reduction of risks posed by contaminated sediment. To receive public input in developing the Strategy, EPA held three national forums on contaminated sediment management [5]. EPA also requested public comments on an outline of the strategy and a draft strategy document [6].

The goals of EPA's Contaminated Sediment Management Strategy are: 1) to develop methodologies for analyzing contaminated sediment so that sediment contamination and associated ecological and human health risks are consistently assessed; 2) to prevent further contamination of sediments that may cause unacceptable ecological or human health risks; 3) when practical, to clean up existing sediment contamination that adversely affects the Nation's waterbodies or their uses, or that causes other significant effects on human health or the environment; and 4) to ensure that sediment dredging and the disposal of dredged material continue to be managed in an environmentally sound manner.

2.   **EXISTING STATUTORY AND REGULATORY
      AUTHORITY TO MANAGE CONTAMINATED
      SEDIMENT**

    EPA has the authority under numerous statutes to address contaminated sediment.   These statutes include: the National Environmental Policy Act (NEPA); the Clean Air Act (CAA); the Coastal Zone Management Act (CZMA); the Federal Insecticide, Fungicide, and Rodenticide Act (FIFRA); the Marine Protection, Research, and Sanctuaries Act (MPRSA); the Resource Conservation and Recovery Act (RCRA); the Toxic Substances Control Act (TSCA); the Clean Water Act (CWA); the Great Lakes Water Quality Agreement of 1978, as amended by protocol signed on November 18, 1987; the Comprehensive Emergency Response, Compensation, and Liability Act (CERCLA); and the Great Lakes Critical Programs Act of 1990.   A complete summary of EPA authorities for addressing sediment contamination is provided in Contaminated Sediments Relevant Statutes and EPA Program Activities [7].

    Many EPA offices implement these statutory authorities or coordinate implementation in specific geographic areas, such as through the Chesapeake Bay Program, the Great Lakes National Program, and the Gulf of Mexico Program.   Depending on statute and program structure, EPA's regional offices and the states may also exercise wide latitude in their determination of sediment quality and environmental impacts.

    Implementation of these programs by different EPA program offices under a wide range of statutory authorities has created some inconsistencies in procedures for assessing the relative risks posed by contaminated sediment, and has increased the potential for duplication in the areas of research, technology development, and field activities.   In order to effectively manage the problem of sediment contamination, EPA must strive to coordinate activities among the Agency's program offices to promote and ensure consistent sediment assessment practices, consistent consideration of risks posed by contaminated sediments, consistent decision-making in managing these risks, and wise use of scarce resources for research, technology development, and field activities.

3. STRATEGY FOR ASSESSING SEDIMENT CONTAMINATION

To implement effective pollution abatement and control programs for contaminants that are accumulating in sediments, and to take appropriate remedial action at sites with identified sediment contamination, EPA has developed a strategy for assessing the extent and severity of sediment contamination. The assessment strategy outlines actions that EPA can take to generate and interpret the environmental data needed to: 1) consistently assess the ecological and human health risks of sediment contaminants and take appropriate regulatory action under the Agency's existing statutory authorities; and 2) identify sites where contaminated sediment remediation is needed, and rank those sites according to the extent and severity of contamination as well as associated ecological and human health risks.

All EPA program offices can use consistent chemical and biological test methods to determine whether sediments are contaminated. EPA is developing standard sediment toxicity test methods to provide high quality data in support of regulatory and enforcement actions for pollution prevention, contaminated sediment remediation, and the management of dredged material disposal. Test methods will be available to address a variety of situations ranging from screening to relatively definitive tests of the effects of contaminated sediments on biological organisms. Test methods will be tiered in order to promote efficient use of resources and screening of sites.

The initial sediment toxicity bioassay species and methods were selected for standardization on the basis of consensus reached at an Agency-wide workshop on tiered testing issues for freshwater and marine sediments held September 16 to 18, 1992 [8]. EPA has recently published the first standard methods for assessing the acute toxicity of sediment-associated contaminants in freshwater, estuarine, and marine systems [9,10].

Pursuant to Sections 304(a)(1) and 118(c)(7)(C) of CWA, EPA is developing sediment quality criteria for the protection of benthic organisms. These criteria are designed to be applied where total organic carbon (TOC) equals or exceeds 0.2% of the sediment dry weight, the primary route of exposure is direct contact with the sediment, and the sediments are continually submerged or there is information indicating that equilibrium has been established between water and sediments. EPA has published and requested public comments on proposed sediment criteria for five chemicals [11]. Comments were requested on documents proposing sediment criteria for the nonionic organic compounds, acenapthene, dieldrin, endrin, fluoranthene, and phenanthrene [12-16]. In addition, comments were requested on technical support documents describing the basis for derivation of the criteria [17,18]. EPA is also developing and will request public comments on a user's guide to application of criteria in watershed protection and ecological risk assessment. The final criteria, user's guide, and technical support documents will be published after considering all public comments. EPA will develop sediment quality criteria for additional nonionic organic compounds using a methodology called the Equilibrium Partitioning Approach (EqP). EPA selected this method after considering a variety of approaches that could be used to assess sediment contamination. A technical review of the criteria and supporting science was conducted by the Science Advisory Board (SAB) [19]. EPA is also developing methodologies for determining sediment quality criteria for metals and ionic organic contaminants. The Agency's SAB has reviewed Office of Water and Office of Research and Development documents describing the application of the Equilibrium Partitioning Approach to five divalent cationic metals: lead, nickel, copper, cadmium and zinc [20,21]. The approach developed by EPA to derive sediment criteria for these metals relates the bioavailability of metals to acid volatile sulfide and organic carbon binding phases in sediment.

In accordance with the requirements of Title V of the Water Resources Development Act of 1992, EPA is conducting a comprehensive national survey of data regarding sediment quality in the United States (Site Inventory). EPA is compiling available information on the quantity, chemical and physical composition, and geographic location of pollutants in sediment. The first biennial report to Congress on the Site Inventory will be completed in 1995. The Site Inventory will be maintained and updated on a regular basis by EPA so that it can be used to assess trends in both sediment quality and the effectiveness of existing regulatory programs at the federal, state, and local levels. The design of the Site Inventory is described in two EPA documents, "Framework for the Development of the National Sediment Inventory," and "Proceedings of the National Sediment Inventory Workshop," [22,23].

EPA has also developed an inventory of sources of sediment contamination (Source Inventory). The Source Inventory is presented in an EPA publication, "National Sediment Contaminant Point Source Inventory: Analysis of Release Data for 1992," [24]. The Source Inventory will be useful to: 1) identify sites where sediment contamination is likely to occur at levels adverse to human health and the environment and undertake sediment monitoring at those sites; 2) target pollution prevention and source control activities by identifying industrial categories contributing sediment contaminants to surface waters; 3) select industries for the development of effluent guidelines on the basis of quantities of toxic sediment contaminants discharged; and 4) target NPDES permitting and enforcement actions to protect sediment quality. The Source Inventory, used in conjunction with the Site Inventory, should allow determinations to be made about whether sites are problematic due to past environmental abuses or as the result of ongoing contamination. EPA will update the Source Inventory every two years.

4.   STRATEGY FOR PREVENTING SEDIMENT CONTAMINATION

Implementation of an effective program to prevent sediment contamination from occurring is the most environmentally protective and, in most cases, cost-effective way to address the problem. EPA's current statutory and regulatory authority is adequate to prevent many sediment contaminants from being released to the environment. The strategy for preventing sediment contamination describes the actions that EPA can take under a number of different statutes, including FIFRA, TSCA, RCRA, CAA, and CWA to prevent sediment contamination.

FIFRA gives EPA the authority to ban or restrict the use of pesticides that have the potential to contaminate sediments, if the risks to "nontarget" organisms are judged to be unreasonable. In making decisions on pesticides, FIFRA requires EPA to consider economic, social, and environmental costs and benefits. Sediment toxicity is not currently addressed in routine test procedures and risk assessments for pesticide registration, reregistration, and special review. Under FIFRA, EPA can require, however, the use of sediment toxicity tests to support registration of pesticides likely to contribute to sediment contamination.

EPA has authority under TSCA to regulate new and existing chemicals that have the potential to contaminate sediments, if the resulting ecological or human health risks are judged to be unreasonable. The Office of Pollution Prevention and Toxics (OPPT) can incorporate into routine chemical review processes, performed under Sections 4 and 5 of TSCA, the assessment of environmental fate and effects of toxic chemicals that could potentially contribute to sediment contamination. OPPT can contribute most significantly to the management of contaminated sediment through its pollution prevention efforts.

5.   STRATEGY FOR ABATING AND CONTROLLING SOURCES OF SEDIMENT CONTAMINATION

The goal of CWA is to restore and maintain the chemical, physical, and biological integrity of the nation's waters. National Pollution Discharge Elimination System (NPDES) permits are the primary means for preventing the discharge of pollutants into water from point sources. Under Sections 301, 304, 306, and 307 of CWA, EPA has set effluent guidelines or minimum, technology-based requirements for municipal dischargers (e.g., primary and secondary treatment standards) and sets similar requirements for industrial dischargers (e.g., best available technology economically achievable and pretreatment standards for existing sources). Under Section 301 of CWA, NPDES permits must also include additional limits as necessary to achieve applicable water quality standards.

Under Section 304(m) of CWA, EPA is required to publish a biennial plan that establishes a schedule for the annual review and revision of promulgated effluent guidelines and identifies categories of sources discharging toxic and nonconventional pollutants for which guidelines have not yet been published.

EPA's effluent guidelines program has evaluated risk as one criterion used to select the types of industries for guidelines development. In the next Effluent Guidelines Plan, the Agency intends to propose adding sediment contamination as a specific factor in the selection of these industry types.

Although in many cases past discharges are partly responsible for today's contaminated sediment problem, sediment quality problems are not solely the legacy of past discharges. Monitoring and assessment data compiled by federal, state, local, and private sources indicate that currently discharging sources may contribute to sediment contamination. Once EPA has developed a standard set of chemical and biological sediment test methods, EPA and the states will be able to use these methods in the process of developing water quality-based permit limits to protect sediment quality for targeted discharges. Toxicity bioassays may be used to confirm whether point source contamination of sediments causes or contributes to aquatic life toxicity. Sediment toxicity identification evaluations can be performed to identify the chemicals causing the toxicity. For human health and wildlife protection, bioaccumulation bioassays can be used to confirm that the chemicals discharged are bioconcentrating in the food chain.

6.   REMEDIATION AND ENFORCEMENT STRATEGY

EPA can take actions directed at remediation of contaminated sediments under the Comprehensive Emergency Response, Compensation and Liability Act (CERCLA), RCRA, CWA, the Rivers and Harbors Act, TSCA, and the Oil Pollution Act of 1990. Where sediments are contaminated to levels that cause ecological harm or pose a risk to human health, EPA can strive to implement whatever remediation strategy will most effectively reduce the risk. In certain circumstances, the best strategy will be to implement pollution prevention measures as well as point and nonpoint source controls to allow natural recovery processes such as biodegradation, chemical degradation, and the deposition of clean sediments to diminish risks associated with the sites. In other cases, active remediation may be necessary. EPA will not proceed with an active clean-up, however, when implementation of the remedial alternative would cause more environmental harm than leaving the contaminants in place. EPA has developed guidance for selecting conventional or innovative remediation techniques for contaminated sediment [25].

7.   STRATEGY FOR DREDGED MATERIAL
MANAGEMENT

Approximately 400 million cubic yards of sediment are dredged from the nation's harbors and waterways each year [26]. Of this amount, some 60 million cubic yards of dredged material is disposed in the ocean at sites regulated under MPRSA [10]. The remaining dredged material is discharged in open water sites, at confined disposal facilities, and for beneficial uses regulated under CWA, as well as upland [26].

The U. S. Army Corps of Engineers (COE), as the federal agency designated to maintain navigable waters, conducts a majority of this dredging and disposal under its Congressionally authorized civil works program [27]. The balance of the dredging and disposal is conducted by a number of local public and private entities. In either case, the disposal is subject to a regulatory program administered by the COE under the above statutes. EPA shares the responsibility of managing dredged material, principally in the development of the environmental criteria by which disposal sites are selected and proposed discharges are evaluated, and in the exercise of its environmental oversight authority. Dredged material management activities are also subject to NEPA, as well as a number of other laws, executive orders, and state and local regulations.

Estimates by the COE indicate that a small percentage of the total annual volume of dredged material disposed, approximately three to 12 million cubic yards, is contaminated to the extent that special handling and/or treatment is required [26]. A number of ongoing and recently completed EPA and COE efforts affect the assessment and management of dredged material, contaminated and otherwise. EPA's Office of Water, in cooperation with the COE, can continue to consistently implement the various statutes and regulations governing dredged material management in an environmentally sound manner. EPA, in cooperation with the Army Corps of Engineers, has completed development of a draft Inland Testing Manual [28] to provide consistent national guidance on testing dredged material for discharge to waters of the United States. A final version of this document will be published in 1995. A similar Ocean Testing Manual is also available [29]. In addition, EPA has recently published a document providing technical guidance on quality assurance and quality control for sampling and chemical analysis of sediments, water, and tissues for dredged material evaluations [30]. The two agencies are also working on a comprehensive ocean disposal site designation, management, and monitoring guidance document.

8.   RESEARCH STRATEGY

EPA's Office of Research and Development (ORD) is engaged in a comprehensive, coordinated program of research that will identify relationships between sediment contaminants and the viability and sustainability of benthic ecosystems, and ultimately will clarify how such information can be used to direct source control and pollution prevention strategies. ORD can support EPA's contaminated sediment management program by undertaking research to develop: 1) methods to assess the ecological and human health effects of sediment contaminants; 2) chemical-specific sediment quality criteria; 3) sediment pollution source allocation methods; and 4) sediment clean-up methods for sites where natural recovery is not appropriate.

9. OUTREACH STRATEGY

Outreach is a critical component of the EPA's Contaminated Sediment Management Strategy. Public understanding of the ecological and human health risks associated with sediment contamination, and of solutions to the problem, is key to successful implementation of this Strategy. EPA can therefore initiate an outreach program in support of Strategy objectives. In implementing the outreach program, EPA can draw upon the experiences of successful public-private partnership programs.

The primary goal of EPA's contaminated sediment outreach program is to educate key audiences about the risks, extent, and severity of contaminated sediments, the role of EPA in solving contaminated sediment problems and the way in which stakeholders will be involved in addressing the problems. EPA's outreach program must therefore have four key elements: 1) defining key Strategy themes or messages; 2) identifying target audiences and needs; 3) developing appropriate outreach materials; 4) providing channels to facilitate two-way communication on Strategy issues.

Four themes of EPA's contaminated sediment outreach program can be conveyed by EPA to target audiences. The first theme is that sediment contamination comes from many sources, which must be identified, and that source control options must be evaluated according to risk reduction potential and effectiveness. The second theme is that sediment contamination poses threats to human health and the environment. The risks must be identified and effectively communicated to the public. Third, sediment contamination can be effectively managed through assessment, prevention, and remediation. And fourth, EPA's strategy for managing contaminated sediment relies on interagency coordination and building alliances with other agencies, industry, and the public.

REFERENCES

1. U.S. EPA, "EPA's Contaminated Sediment Management Strategy," EPA-823-R-94-001 U.S. EPA, (Washington, DC 1994).

2. U.S. EPA, "National Perspective on Sediment Quality," Report prepared by Battelle under EPA Contract #68-01-6986 for Office of Water, Criteria and Standards Division, (Washington, DC 1985).

3. U.S. EPA, *An Overview of Sediment Quality in the United States* EPA-905-9-88-002 U.S. EPA, (Washington, DC 1988).

4. National Academy of Sciences, *Contaminated Marine Sediments-Assessment and Remediation*, National Academy Press, (Washington, DC 1989).

5. U.S. EPA, *Proc. EPA's Contaminated Sediment Management Strategy Forums*, EPA-823-R-92-007 U.S. EPA, (Washington, DC 1992).

6. *Federal Register*, 59(167), August 30, 1994.

7. U.S. EPA, "Contaminated Sediments-Relevant Statutes and EPA Program Activities," EPA-506-6-90-003 U.S. EPA, (Washington, DC 1990).

8. U.S. EPA, "Tiered Testing Issues for Freshwater and Marine Sediments," in *Proc. September 16-18, 1992, Washington, D.C. Office of Water, Office of Science and Technology, and Office of Research and Development*, EPA-823-R-93-001 U.S. EPA, (Washington, DC 1992).

9. U.S. EPA, "Methods for Measuring the Toxicity and Bioaccumulation of Sediment-associated Contaminants with Freshwater Invertebrates," EPA-600-R-94-024 U.S. EPA, (Washington, DC 1994).

10. U.S. EPA, "Methods for Measuring the Toxicity and Bioaccumulation of Sediment-associated Contaminants with Estuarine and Marine Amphipods," EPA-600-R-94-025 U.S. EPA, (Washington, DC 1994).

11. *Federal Register*, 59(11), January 18, 1994.

12. U.S. EPA, "Sediment Quality Criteria for the Protection of Benthic Organisms: Acenapthene," EPA-882-R-93-013 U.S. EPA, (Washington, DC 1993).

13. U.S. EPA, "Sediment Quality Criteria for the Protection of Benthic Organisms: Dieldrin," EPA-882-R-93-015 U.S. EPA, (Washington, DC 1993).

14. U.S. EPA, "Sediment Quality Criteria for the Protection of Benthic Organisms: Endrin," EPA-882-R-93-016 U.S. EPA, (Washington, DC 1993).

15. U.S. EPA, "Sediment Quality Criteria for the Protection of Benthic Organisms: Fluoranthene," EPA-882-R-93-012 U.S. EPA, (Washington, DC 1993).

16. U.S. EPA, "Sediment Quality Criteria for the Protection of Benthic Organisms: Phenanthrene," EPA-882-R-93-014 U.S. EPA, (Washington, DC 1993).

17. U.S. EPA, "Technical Basis for Deriving Sediment Quality Criteria for Nonionic Organic Contaminants for the Protection of Benthic Organisms Using Equilibrium Partitioning," EPA-822-R-93-011 U.S. EPA, (Washington, DC 1993).

18. U.S. EPA, "Guidelines for Deriving Site-specific Sediment Quality Criteria for the Protection of Benthic Organisms," EPA-822-R-93-017 U.S. EPA, (Washington, DC 1993).

19. U.S. EPA, "Review of Sediment Criteria Development Methodology for Non-Ionic Organics," in *An SAB Report*, Sediment Quality Subcommittee of the Ecological Processes and Effects Committee, EPA-SAB-EPEC-93-002 U.S. EPA, (Washington, DC 1992).

20. U.S. EPA, "Briefing Report to the Science Advisory Board on the Equilibrium Partitioning Approach to Predicting Metal Bioavailability in Sediments and the Derivation of Sediment Quality Criteria for Metals," EPA-822-D-94-002 U.S. EPA, (Washington, DC 1994).

21. U.S. EPA, "Supporting Publications to Accompany the Briefing Report to the EPA Science Advisory Board on the Equilibrium Partitioning Approach to Predicting Metal Bioavailability in Sediments and the Derivation of Sediment Quality Criteria for Metals," EPA-822-D-94-003 U.S. EPA, (Washington, DC 1994).

22. U.S. EPA, "Framework for the Development of the National Sediment Inventory," EPA-823-R-94-003 U.S. EPA, (Washington, DC 1994).

23. U.S. EPA, "Proceedings of the National Sediment Inventory Workshop," EPA-823-R-94-002 U.S. EPA, (Washington, DC 1994).

24. U.S. EPA, "National Sediment Contaminant Point Source Inventory: Analysis of Release Data for 1992," EPA-823-R-95-006 U.S. EPA, (Washington, DC 1995).

25. U.S. EPA, "Selecting Remediation Techniques for Contaminated Sediment," EPA-823-B-93-001 U.S. EPA, (Washington, DC 1993).

26. Lee, C.R, "U.S. Army Corps of Engineers National Dredging Program. Presentation to the CSMS Forum on; The Extent and Severity of Contaminated Sediments," in *Proc. EPA's Contaminated Sediment Management Strategy Forums*, EPA-823-R-92-007 September 1992, U.S. EPA, (Washington, DC 1992).

27. Moore, D. and J. Wilson, "Presentation to the CSMS Forum on Building Alliances Among Federal, State, and Local Agencies to Address the National Problem of Contaminated Sediments," in *Proc. EPA's Contaminated Sediment Management Strategy Forums*, EPA-823-R-92-007 September 1992, U.S. EPA, (Washington, DC 1992).

28. U.S. EPA and U.S. ACE, "Evaluation of Dredged Material Proposed for Discharge in Waters of the U. S. - Testing Manual (Draft)," EPA-823-B-94-002 U.S. EPA, (Washington, DC 1994).

29. U.S. EPA and U.S. ACE, "Evaluation of Dredged Material Proposed for Ocean Disposal - Testing Manual," EPA-503-8-91-001 U.S. EPA and U.S. ACE, (Washington, DC 1991).

30. U.S. EPA, "QA/QC Guidance for Sampling and Analysis of Sediments, Water, and Tissues for Dredged Material Evaluation," EPA-823-B-95-001 U.S. EPA, (Washington, DC 1995).